RENEWALS 458-4574
DATE DUE

**WITHDRAWN**
UTSA Libraries

# A Prussian Observes the American Civil War

# SHADES OF BLUE AND GRAY SERIES
**Edited by Herman Hattaway and Jon L. Wakelyn**

The Shades of Blue and Gray Series offers Civil War studies for the modern reader—Civil War buff and scholar alike. Military history today addresses the relationship between society and warfare. Thus biographies and thematic studies that deal with civilians, soldiers, and political leaders are increasingly important to a larger public. This series includes books that will appeal to Civil War Roundtable groups, individuals, libraries, and academics with a special interest in this era of American history.

# A Prussian Observes
# The American Civil War

The Military Studies of Justus Scheibert

Translated and edited
by Frederic Trautmann

University of Missouri Press
Columbia and London

Copyright © 2001 by
The Curators of the University of Missouri
University of Missouri Press, Columbia, Missouri 65201
Printed and bound in the United States of America
All rights reserved
5 4 3 2 1   05 04 03 02 01

Library of Congress Cataloging-in-Publicaton Data

Scheibert, J. (Justus), 1831–1903.
  [Bürgerkrieg in den Nordamerikanischen Staaten. English]
  A Prussian observes the American Civil War : the military studies of Justus Scheibert / translated and edited by Frederic Trautmann.
      p.   cm.—(Shades of blue and gray series)
  Originally published in German under titles: Der Bürgerkrieg in den Nordamerikanischen Staaten; and, Das Zusammenwirken der Armee und Marine.
  Includes bibliographical references (p.   ) and index.
  Contents: pt. 1. The sinews of war—pt. 2. The sinews applied.
  ISBN 0-8262-1348-0 (alk. paper)
  1. United States—History—Civil War, 1861–1865—Campaigns.
  2. United States. Army—Drill and tactics—History—19th century.
  3. Confederate States of America. Army—Drill and tactics—History.
  4. Military art and science—United States—History—19th century.
  5. Military art and science—Confederate States of America.  6. Scheibert, J. (Justus), 1831–1903.  7. United States—History—Civil War, 1861–1865—Personal narratives, Prussian.  I. Trautmann, Frederic, 1936–
II. Scheibert, J. (Justus), 1831–1903. Zusammenwirken der Armee und Marine. English.   III. Title.   IV. Series.
E470 .S313 2001
973.7'3—dc21                                                                                          2001027552

∞™ This paper meets the requirements of the
American National Standard for Permanence of Paper
for Printed Library Materials, Z39.48, 1984.

Text designer: Elizabeth K. Young
Jacket designer: Vickie Kersey DuBois
Typesetter: BOOKCOMP, Inc.
Printer and binder: The Maple-Vail Book Manufacturing Group
Typefaces: ITC Veljovic, Bodoni Condensed

# Dedication

Again to the great translators past, present, and to come and, this time, especially

Jerome
Hanayn ibn Ishaq
Thabit ibn Qurra
Sir Richard Francis Burton
Eleanor Marx
H. T. Lowe-Porter
Samuel Noah Kramer
Eden and Cedar Paul
Richard and Clara Winston
Ralph Manheim
Anthony Burgess
Burton Raffel
Gregory Rabassa
William Weaver
Margaret Sayers Peden
Leila Vennewitz
Seamus Heaney
Christopher Middleton

*Each [wise man] has answered [the questions that disturb and puzzle and confound us] according to his ability, by his word and his life.*  —*Thoreau,* Walden

*Count the American Civil War one of the world's historic wars.*
 —*Scheibert,* infra, *chapter 6*

*The American war probably excels as the textbook of warfare.*
 —*Scheibert, chapter 1*

*Perhaps no other episode [than the Mississippi campaign] provides so many and such good examples [of combined operations in modern war].*  —*Scheibert, chapter 10*

*Believe me: I saw it.*  —*Scheibert, chapter 5*

*You poor mad things—what will become of you?*
 —*John Ruskin to Charles Eliot Norton, February 1, 1862.*

*Soon we shall go out of the house and into the convulsion of the world, out of history and into History and the awful responsibility of Time.*
 —*Robert Penn Warren,* All the King's Men

# Contents

Acknowledgments  ix
Preface: Editor-Translator's Notes  xi
To the Reader: Maps, Text, Editing and Translating  xiii
Author's [Scheibert's] Notes  xv

Prologue: The Great Captains  1

## PART ONE   THE SINEWS OF WAR

1. Strategy  13
2. Tactics  33
3. Infantry  53
4. Cavalry  59
5. Artillery  79
6. Engineering  92
7. Medical Service  105
8. Navies  117

## PART TWO   THE SINEWS APPLIED: A SOLDIER'S HISTORY

9. War All the Way: Mostly East  135
10. War in the West: Combined Operations  151

Epilogue: The Lesson of the War  192
Appendix: The War in the West  201
Notes  209
Bibliography  227
Index  239

# Acknowledgments

*Thanks from the bottom of my heart to*
Cathy Meaney and her staff of the Interlibrary Loan Unit, Temple University Libraries: again they brought so much, so far, so fast;
John C. Van Horne's staff of the Library Company of Philadelphia: again they shared their treasures and enriched my book;
David M. Neigher: the man of wisdom, sense, and good judgment;
Beth Trautmann: she helped;
Mary Dunn and staff, Information Processing Center, Temple University: the rest of the world should know its stuff and work so well and be so congenial;
Sara Davis, University of Missouri Press: a fine editor; and
Lady Luck, her especially, the best of muses; she smiled again and again; she was a lady day and night.

# Preface: Editor-Translator's Notes

The American Civil War meant much to Europeans: to their commerce, diplomacy, statecraft, balance of power, and military affairs. European armies could learn from American examples. All the major newspapers of Europe sent correspondents; every government, agents and operatives; and armies, observers, to learn tactics, strategy, and equipment. "The Americans have something cooking, that ingenious, shrewd, inventive Yankee race. Something big. Let's get in on it. They'll solve the new problems. Could be a feast of answers."

To learn the military lessons, and what to teach its army, Prussia sent one of its military's brightest students and strongest teachers, Justus Scheibert (1831–1903), captain (later major) of engineers, career officer of distinction, authority on military matters, editor, author, and lecturer. He spent seven months in the Confederacy in the decisive year of 1863. Robert E. Lee liked and respected Scheibert so much that he took him into his headquarters and gave him shelter, food, information, and friendship. Scheibert to the end of his life counted Lee and many other Confederate generals friends. In Lee's camp, other foreign observers found hospitality elsewhere if at all.

Scheibert took part in the battles of Chancellorsville, Brandy Station, and Gettysburg; experienced Charleston under siege and bombardment; observed every arm of the Confederate service; inspected fortifications and entrenchments; talked to eyewitnesses about what he had not seen himself, including any detail about everything on the Union side; and ran the blockade—all with notebook in hand, pencil poised. He saw and felt our war.

Back in Prussia, notebook still in hand, he applied his research to Prussia's world-class libraries. Scheibert tells in his memoirs that Bismarck himself asked Scheibert to guide him through the works captured from Denmark at Düppel in 1864. He wrote out what he learned, for an audience that knew little, if anything, about America or its war. "The basics, Scheibert, down to brass tacks." What

Scheibert learned, taught, and wrote informed Prussia in three wars, and Germany in two.

Scheibert's books, though manuals for Prussian and German officers, became handbooks and guides to our war's military side. They remain among the best of the kind. The one on action along the Mississippi gave the officers the first study of combined operations anywhere ever. Time has rated his works with the strong, solid military histories of our war. What book has bettered them on their subjects, or what articles or essays, and by an eyewitness? Indeed, little noticed and long forgotten, hidden by the German language, they read now like a new, fresh, bold interpretation of our war, and like Scheibert's criteria for the commander, realized to perfection in Lee: intelligent, flexible, incisive, brave, and vigorous if not passionate of resolve and of action. Who later dare quarrel with him who had been on the spot and seen the war, or doubt his years of library research?

Until now, only readers of German—or for parts, French in translation—could know what Scheibert learned about the military details of our war. Now, here, in one work, I have brought his two together for the most deserving audience, the one that should have had them all along: Americans—translated into our language, edited for organization and with annotations and explanatory notes, and adding a bibliography of evidence for reference, documentation, and further reading.

Scheibert can speak for himself here to you. Let him. Hear him. Heed him.

**Frederic Trautmann**
Glenside, Pennsylvania
summer, 2000

# To the Reader: Maps, Texts, Editing and Translating

**K**eep modern maps and battle diagrams at hand. Scheibert's, however well intentioned then and there, have been omitted as gratuitous here and now: roughly drawn, inaccurate, too few, in German, and the maps (done before metric had become the European standard) drawn to the scale of German miles. Scheibert's few would have necessitated more anyhow. The bibliography offers a list.

## Texts

Study One first published as *Der Bürgerkrieg in den Nordamerikanischen Staaten: Militärisch beleuchetet fur den deutschen Offizier*. Berlin: Ernst Siegfried Mittler und Sohn, Konigliche Hofbuchhandlung, 1874.

Study Two first published as *Das Zusammenwirken der Armee und Marine. Eine Studie illustriert durch den Kampf um den Mississippi, 1861-1863*. Rathenow: Verlag von Max Babenzien, n. d. [1887?]

Neither has appeared in English before. Contrary to Douglas Southall Freeman in *R. E. Lee* (4:560), Study One has never been published in English translation. He cites it in German but says it has been translated into English. Like every other author in English he does not mention Study Two. Probably none of them knew it existed.

## Editing and Translating

Editing has been diverse: no oftener and no heavier than necessary, but as often and heavy as necessary, therefore seldom and light in some places, heavier and oftener in others. Not the sledgehammer and the pick and shovel, but the jeweler's mallet and the dainty ladies' comb and brush.

Accordingly, making two books into one and organizing for clarity and logic have demanded rearrangement to articulate them. The smaller, on combined operations, has gone mainly into chapter 10 here. The larger, on the entire war, encloses it. The summary of the larger, at the start of the German version, has been moved to the end, and the commanders' biographies from there to the start. Small reorderings within have provided transitions as well as clarifying and emphasizing as needed. Names, dates, ranks, and places have been added without notice to the reader, where vagueness or gaps called for them. In the study of combined operations, intermediate summaries have been delayed until the summing up. Two chapters of the large book, "Infantry and Tactics" and "Artillery and Engineering," have been made four, "Infantry," "Tactics," "Artillery," and "Engineering." Sizes, calibers, and distances have been left as Scheibert put them in metric. His German mile—as archaic and specialized now as the furlong, hand, and league in English—have been converted at 4.5:1, German to statute. Artillery ranges have been left as pace (*Schritt*), about a yard, still much used by old hands in the field artillery. Most of Scheibert's notes have been moved without notice into the text; those retained as notes, labeled as his. Nearly all, being the editor-translator's, have been added for reasons evident with each. These have been kept to the minimum, to let the eyewitness and the expert speak for themselves in Scheibert's words.

In translating, everything Scheibert said has been kept, not in word-for-word transcription (impossible anyway), but in ideas as they would have been expressed had Scheibert been Trautmann writing in English.

# Author's [Scheibert's] Notes

*Sine ira et studio.*

That motto, "Neither partial nor in anger," should head any work claimed factual and objective. Especially must I put it here. You, friends and associates, know my partiality and my passion in that war—I fought for the South and believed in it body and soul. Partiality aside here and passion calmed now, I have striven for a book of the military and for the military, by a military man, and neither partial nor in anger. To you, colleagues, I swear a dispassionate critique.

In that frame of mind I have studied North and South, learned from each, and praised and censured both. I acknowledge Northern superiority in medical service, artillery, and the navy; and on them I refer mostly to the North. On infantry, cavalry, and strategy, the South has more to teach. Thus if I take opposite viewpoints in respective chapters, you will not fault me for shifting the emphasis to keep to my purpose. I want above all to teach you the lessons of that war. In the prime example, Lee managed four years' resistance to the power of the North while one Northern commander after another made mistake upon mistake and aided his dogged resistance. Rather than dilate upon them, however, I shall show you how he prevailed. His model will teach any soldier more than would a dissertation on their mistakes.

The author
Minden, June 1874

I acknowledge colleagues of the Imperial Navy, who encouraged me to start the study of combined operations. The Navy's generous support helped me to finish it.

**The author**

# A Prussian Observes the American Civil War

# Prologue
## The Great Captains

*The genius of a man grew and glittered.*

### J. E. B. Stuart–Stonewall Jackson–W. T. Sherman–U. S. Grant–R. E. Lee

Let me begin with the foremost commanders. Space does not permit portraits, only sketches. Meeting the main characters now, albeit briefly, we can better appreciate their roles in the drama that follows. Lee I knew personally, as well as Jackson and Stuart. The famous cavalry leader Stuart, antenna to his army, met the enemy first. Let him take the lead here. Von Borcke's pen brings him to life in a minimum of strokes.[1]

"Colonel Lee pointed out to me a man, galloping rapidly along on an active, handsome horse, as Stuart, the man whose arrival I awaited so anxiously, and who subsequently became to me one of the truest and best friends I have had in this world.

"General Stuart was a stoutly built man, rather above the middle height, of a most frank and winning expression, the lower part of his fine face covered with a thick brown beard, which flowed over this breast. His eye was quick and piercing and of a light blue in repose but changing to a darker tinge under high excitement. His movements were alert, his observation keen and rapid, and his person the sum of vitality guided by instinct. Because he could therefore act fast and be right without pausing to think, he was to me the model of the dashing cavalry leader. Before the breaking [out] of hostilities between the North and South, he had served in the First United States Cavalry, of which regiment General Joseph E. Johnston was the Lieut. Colonel, against the Indians of the Far West, and was severely wounded in an encounter with the Cheyennes on the Solomon's Fork of the Kansas River, in July 1857. In that wild life of the prairie, now chasing the buffalo, now pursuing the treacherous savage, whose steeds were almost as fleet as the wind, Stuart had passed nearly all his waking hours in the saddle, and thus become one of the most fearless and dexterous horsemen in America, and he had acquired a love of adventure which made activity a necessity of his being. He delighted in the neighing of the chargers and the roll of the bugle, and he had something of Murat's weakness for the ostentation of the parade. He betrayed this latter quality in his jaunty uniform, which consisted of a small grey jacket, trousers of the same stuff, and over them high military boots, a yellow silk sash and a grey slouch hat, surmounted by a sweeping black ostrich plume. Thus attired, sitting gracefully on his fine horse, he did not fail to attract the notice and admiration of all who saw him ride along. This is not the place to submit any remarks on the military character of General Stuart. His deeds will form the most considerable portion of this humble narrative, and out of them my estimate of his soldierly qualities will naturally arise."

## 2

Stonewall, the original, how he tempts a biography definitive and elaborate! How hard to be satisfied with these few strokes! How they compromise that short yet abundant and heroic life! Yet neither you nor anyone else will see the distinguished general in how he began or the way he looked, except that his parents intended him early for a military career. He grew to suggest the scholar, not the soldier: moderate height, stooped and awkward of gait, with eyes dark and intelligent, and face framed by a black beard. If the long raven hair, weak voice, and threadbare clothes did not suggest, let alone become, the soldier, what of the horse and horsemanship? No great rider, he brought little elegance to a brown that had degenerated to a bony nag. West Point graduated him in artillery, a second lieutenant of gun rather than horse. Excellence against Mexico in 1847 soon won him promotion to first lieutenant and captain at Contrera and Churubusco, and major at Chepultepec.

The Mexican climate damaged his health; he had to resign from the army in 1852. He joined the faculty of the Virginia Military Institute, where he became known as a crank, the pedantic, rigorous, stern professor who put students off. Nothing but originality suggested a great commander. Students disliked him; he seemed to lack everything but eccentricity and grew to be a fixture for caricature, winning many a nickname, including "old Tom." As it had taken the Mexican to awaken greatness and bring forth excellence, so it took the

Civil War to renew and enlarge them. Otherwise he would likely have lived out his life as another of the countless souls known perhaps for being odd, certainly for nothing else.

In 1861, at age thirty-seven, he accepted the call to Harpers Ferry to head a small observation unit.[2] From the start we shall see that he possessed the talent for command. He could train troops—he could lay and carry out plans—but we say enough about his superiority in the pages that follow; we need not start here. It suffices that we stress an aspect of his style now. A Presbyterian like Cromwell, he led men not only as soldiers, he the general; but also as congregation, he the pastor. True and devoted to the faith, pious, childlike in sincerity, he entertained no idea without testing it in prayer, fought no battle without seeking the blessing of God. He prayed before battle. His servant, a Negro, bore witness. "Massah, morning big battle, prays so much."[3]

He grew strong by the struggles of the spirit and the lifting of the soul to God. Fidelity to service and passion to effect his ideas came of abandoning himself to God's will. Thereby he manifested the confidence that neared elation and shone in his face when mounted at the head of a campaign. Thus he evinced the character that won the devotion of the troops. In their eyes he became perfection. They loved, they revered him. So he served and led with such force that must trample everything mediocre.

He died as he lived: the faithful believer, the heroic soldier. "A. P. Hill," he said (his last words), "get ready to fight."[4] His death caused throughout the army the grief and mourning only some could appreciate who saw it. I saw it. The loss shook none more than Lee, a Blücher who lost his Gneisenau.[5] When Lee learned of the amputation of Jackson's arm, Lee said: "Yes, pity him, losing his left arm. Pity me all the more. I have lost my right." Lee's letter, when he heard Jackson had been hurt, testifies best to what the two felt for each other. "I should have chosen for the good of the country to be disabled in your stead."[6]

As long as men write the history of war they will remember Jackson, and every soldier's heart lift at the heroism of the brave, pious general.

### 3

I never met General Sherman. I use for my sketch Major von Meerheimb's excellent *Sherman's Campaign in Georgia*.[7]

## 4  A Prussian Observes the American Civil War

"William Tecumseh Sherman, born in Ohio in 1820, came of an English family long in America: since the 1600s. His father, an attorney, died young. Sherman, left without an inheritance, went to West Point at the nation's expense. The artillery lieutenant served several years in Florida in miserable, thankless Indian fighting. He took his discharge in 1851 and became a businessman, running a wholesale house in San Francisco, until a few years later the state of Louisiana installed him as the superintendent of a military school. Secession rushed darker and darker forward in 1860 and 1861. He saw that the sides could not be reconciled. They would split—soon.

"Already in a semimilitary post, he left it and put himself at the disposal of the government in Washington. A regiment of volunteers elected him colonel. Later he took over a brigade. He distinguished himself at First Bull Run, then Shiloh. He moved up to a division, then a corps, then the Army of Tennessee in the Mississippi campaign in 1863, and then took an essential part in the victory at Missionary Ridge. Meanwhile in those years he had been laying his plans and studying the terrain for a thrust into the heart of the South. He would destroy the railroads and all communications, isolate the Southern forces, and cut off their subsistence.

"Brilliantly he carried through the campaign to Atlanta against Johnston and Hood. Atlanta taken, he commenced his world-famous campaign to Savannah, captured it and Charleston, and marched north as if to encircle Lee at Petersburg. Lee had defended his positions a long time; now he must surrender to Grant. The war ended. Sherman, advancing from the south, had been decisive.

"He would seem by energy and intelligence one of America's leading generals. He laid good plans, down to the last detail, and carried them out with equal flair. Reckless energy and ruthless drive propelled him until they hurt and beyond hurt. Appearance corresponded to personality: almost six feet tall, lean, sinewy rather than muscular, tough, resilient, of iron constitution, healthy as a horse, and able to endure any hardship. Features fiercely drawn, restless piercing eyes beneath beetling brows; forehead furrowed but nonetheless assertive: they registered a nature ruled by will and intellect—mostly will.

"He enjoyed in full measure the love and trust of his officers and men; for—serious, self-sacrificing, tireless, ever active, caring little for himself—he cared much for his troops. Movingly they showed what they felt for this supposedly cold, off-putting man, as on the occasion of the death of a child he loved. Compare Grant, Sherman's friend from start to finish. A man of imperturbable, phlegmatic equanimity, firm, steadfast, never ruffled, Grant let nothing upset his balance. Emotion, the nervous element, ran Sherman, the man never calm, ever agitated, always on the move in body and soul.

"His correspondence shows a piercing, superior mind. Let me give a few examples to show what manner of man we have here. Many in Europe in comfortable indolence customarily consider them [the Americans] barbarians or 'money makers.' Grant and Sherman illustrate two distinct types of the Anglo-American character.

"Does the terrible waging of war, the deliberate laying waste of rich, lush regions, remind us of the Thirty Years War? Balance it with the fact that Sherman took such measures with a heavy heart and only because they alone would bring a sure and speedy end to the war. He and all good people in the army and among civilians wanted the war ended; they longed for the end. He had just taken Goldsborough, he stood at the height of his fame, when he wrote that he had enough of war. He compared its glitter and glory to the deceptiveness of moonbeams. He deplored success reaped after the sowing of misery and the spilling of blood. He knew the stakes: save the Union or be obliterated. Crush secession. But, enemy beaten, he would say to each and every rebel: 'Go and sin no more.' So he said and so he acted. Surrender came on terms easy enough for the Confederate states and army; and he did what he could to ease their fate during the peace that followed, and to see their rights honored."

## 4

We turn from Sherman to the officer closest to him, Ulysses Simpson Grant, now president. Obstruction after hindrance upon obstacle confronted Grant. In the end his plans, quickly made and skillfully drawn, beat the South because he, flexible by nature, remained so despite everything against him and in the face of setback upon loss after defeat. Subordinates trusted him for being open and direct. The correspondence shows it: his frankness in every word of his, their trust nowhere clearer and plainer than in Sherman's vital, artful letters. Grant knew how to box the compass and set the course. With the energy that brought him fame he did things his way, not as the

Lincoln administration might have done. Instinct for true north and the homing sense avoided sidetracks, guided his policies to victory and the general good, and carried him to the presidency itself.

He had entered the regular army and served with distinction in Mexico; then left the army, reasons not clear. The ex-captain lived in St. Louis in less-than-agreeable circumstances. He tried this occupation and that line: wool handler, bill collector, auctioneer, realtor. He declared bankruptcy; he had failed at everything.[8] He fled at last to Galena, Illinois, and refuge behind the counter of his father's leather-goods store. Nothing about the clerk suggested he would soon determine the fate of a continent.

War broke out. He organized the Twenty-first Illinois and became its head in June 1861. By September, already promoted to brigadier general, he commanded two brigades. Frémont helped him to the top post of the District of Southeastern Missouri. Strong leadership distinguished him against Polk in the West.[9] Nota bene, in a contest of unfit troops on either side, Grant would take the offensive when all seemed lost, cow the foe, and often win. We will see his drives of power and skill in Forts Henry and Donelson; how, penetrating Confederate lines at the center, he kept Missouri, Kentucky, and northern Tennessee for the Union. He took Vicksburg with the same mettle in 1863—likewise Chattanooga—therefore became general in chief of the armies of the United States. Then his talent failed against the genius of his counterpart. Only with Sherman's aid could he fight to a successful end.

Prevailing over a narrow-minded cabal, he treated generously his noble and awe-inspiring opponent and arranged and accepted a gracious surrender. He deserved the seat of national honor. Later he took it as head of the Republic.[10]

## 5

I address this book to my fellow officers and other readers with military interests in this and all wars. Therefore I wanted to look at the facts as objectively as possible and submit an unbiased account, my interpretation of the military science. I tried. But the genius of a man grew and glittered. I concerned myself with his military side, of course. My testimony, like the rest by those who knew and studied him, awards every laurel, not because I or they intended to, but because he deserved it. His nation and army passed an even broader judgment: "Great as a soldier, greater as a man!" So he should end and must crown these sketches, be it only as the soldier.

I saw him at fifty-eight in 1863: imposing, grand, elegant in the saddle, a military presence to the nines and in spades. The war had whitened the frame to the strong, noble face. In it glowed beautiful eyes, frank and almost coal-black. He spoke little of a self that could be summed up in a word, *calm*.

His old English family had left Britain at the time of the revolution. Marrying a granddaughter of George Washington, he came into the first president's estates. Before the war he counted among America's leading political figures.[11]

West Point, officer of engineers, Mexican service with honor: in sorrow he the patriot watched the fires of civil war ignite. He flayed those who set them. Never had he followed inclination or desire; always listened for the call of duty, his guide to right conduct, and obeyed what he heard. The homeland Virginia joined the Confederacy; his indecision ended. He obeyed the call again. The North offered—he foreswore—a bright future.[12] He cut the ties and quit the relationships of long, respectable national service. He accepted as his own the homeland's misfortunes, helped bear them with honor, until he and it collapsed under their weight. He had served the Union; it had acknowledged and rewarded him. But now, born to the South, he had made the hard choice and would live with it.

What did he give during the war, if neither hardware nor firepower? Only once, soon after Jackson's death, did we see the lapse of leadership on his part. At Gettysburg the loss of the good right arm, a wound too fresh, crippled the army and impaired Lee. Vigor and suppleness would not return until the wound healed. His gifts nevertheless astound the attentive observer. Generalship proved Lee the virtuoso of offense, maestro of defense, and prodigy of anticipating the enemy's intent. What largesse of the sinews of war: plans, strategy, and the ways to deploy troops! He applied with bravura the infantry tactics he originated. Cavalry, realizing under him its strategic value, took on the role that brought renown. Artillerists often praised his understanding of their arms. Seldom did he err in choice of position, even to the enemy's wonder and praise. Seldom therefore did he draw the sword, but he sheathed it as victor.

He did not bother with details outside his province as general in chief. In it he neglected nothing and toiled at what concerned him. He exemplified deliberateness, and deliberation informed his decisions. A cavalry general must plan at the speed of light. Our general in chief, heading an army and obligated to a nation, mulled and labored over plans, crafting, comparing, weighing, balancing. Responsibility devolved on him. He felt it. In him the nation had put every trust. He respected it. In his hands he saw the flower of the youth, the bloom of hope, and the fate of the nation itself. He believed he must answer to the people and to God. Fidelity drove him to consider the chance and avoid the mishap that could foul his calculations and wreck the enterprise. Then he would devote himself to carrying out as fast as possible the plans so painstakingly drawn. He said to me after a battle: "I make the plan as good as the power of the human being can make it. But on the day we fight I trust the army's fate to God. My generals must do their duty then."[13]

Meanwhile he would be where able to do *his* duty, not only placing himself to look on, but even to join in the battle. Fire and storm could not budge him. In May of 1863 at Chancellorsville his headquarters lay in a battery's field, but from May 4 to 7 he stayed put. Some twenty bits of shrapnel and other missiles pierced his tent. Yet he would not move a little sideways and off Fairview Hill. He did go a few paces with us and into a ditch when the battery began to zero in.

Benevolent majesty filled that face. The spacious character contained humanity and kindness without sentimentality on the one hand, and draconian severity cold-blooded but without cruelty on the other. Sympathy at a wounded man would flash in features otherwise unruffled and imperturbable in victory as in defeat. A cloud gathered there but once. In April of 1865 a provision train on wrong orders, a dereliction of duty, left his loyal band hungry and ready to surrender.[14] Humanity and kindness yielded to severity then. Yet, a plan drawn, he never hesitated to commit troops with force, recklessly, to carry it out.

Still his every act bespoke dignity and embodied nobility; he had grown up in the luxury and sophistication of one of the South's wealthiest and most aristocratic families. Yet he set for the army the example of simplicity, being content with the minimum of baggage, scarcely the essentials. This general lived in the plain and severe style of a private. In the field and during winter's months of quarters, unless ill, he never hesitated nor needed to be invited to sleep among his army and out-of-tent. Devotees sent him refreshments, notably wine. He forwarded every bottle to his hospitals.

He headquartered in a mere six to eight tents, little sheets, scant shelter to him, his staff, and their work. In that cramped space they found room for me! I ought not be blamed for anything like pride. True, I had served an apprenticeship with Stuart. But many foreign officers followed the army as observers. Any of them could have been the one to join Lee.

The Confederacy collapsed. "In my prime I led to death one generation of my country's youth," Lee said. "In age I shall atone by educating another." He became director of the academy at Lexington.[15] It turned from ruins into one of the South's most desirable schools. Every Southern student wanted an education at the knee of the old, gray captain.

In the autumn of 1870 a stroke felled him as he said grace. A few days later, October 12, he died with the battle cry on his lips. So ended an original, a paragon of simple, natural intellect, as devoted to duty as to God, hence one of the century's foremost generals.

I can suggest only one as his equal. I mean the first to do his duty—therefore the first soldier of his country—the venerable commander—could I but name him! Out of respect for the living, by my loyalty as an officer, I must not.[16] I wish I might draw the comparison! It would be the capstone to this book!

PART ONE

# THE SINEWS OF WAR

# 1

# STRATEGY

*Union and Confederacy played the absorbing games on the Virginia board.*

The Virginia Theater—Railroad and Telegraph—Food Supply—Leadership in the First Stage—Union Strategy—Confederate—Interior Lines: Minor Instances; Jackson's Shenandoah Campaign; Campaigns against Pope, Burnside, Hooker—Chancellorsville—Lee's Offensive into Pennsylvania—Championship Match: Grant versus Lee—Conclusion

The student of warfare needs examples: communications (railway systems, the telegraph, rapid transport), living off the land and from the storehouse, systems of requisition, and the coordination and cooperation of masses of troops—these aspects of combat as they occurred under the gun. Recent European wars offer them and others, as well as complications among them. Unfortunately, only the officer accomplished in military science will appreciate what they mean for strategy; their convolutions will baffle the beginner. On the other hand, the American war probably excels as the textbook of warfare in discrete and basic forms. The simple, clear circumstances afford instance upon prime instance in the management of units large and small. (In that sentence I speak to the novice.) Even the expert finds the campaigns instructive where we see warfare on the broader canvas, and when we witness the virtuosity of this commander or that strategist. (I speak of Virginia in the last two sentences.) The West avers eloquent models and splendid leaders: Sherman versus Johnson in 1864, say, and their dynamic, edifying maneuvers. But Union and Confederacy played the absorbing games on the Virginia board. The student must know Virginia for that reason. Hence this sketch.

That eastern state, on the coast below the Potomac, asserts level, swampy lowlands oceanside, rolling piedmont inland, and mountains covering the one-third of the state beyond. These parallel ranges of the Alleghenies approximate the angle of the coastline southwest to northeast. The Potomac pierces them at right angles along the northern border. The easternmost range or Blue Ridge gives rise to rivers and creeks perpendicular to the coast and therefore parallel to each other and pouring into the ocean the currents that gashed the rocky soil and carved steep banks. Mostly they must be forded—a map will show the few bridges—but allow it only in dry weather. Communication stalls when rain bloats them and they act up, the rowdies! Nearer the ocean they permit navigation; they widen to funnel-shaped bays, the deep dents that let enemy fleets inland. Countless smaller streams, also steep-banked, run through the level countryside and feed the larger. Forests cover two-thirds of the state. Farms amid forests create their version of oases, isolated refuges in the wooded wilderness.

The few roads, even major routes, asserted the condition that had become proverbial: miserable. In bad weather, for sure in winter, they permitted no communication, none. A few "plank roads" crisscrossed vast, forested reaches with their three-to-four-inch members nailed to stout stringers in design and materials peculiar to the region.[1] Rather than torture themselves on such golgothas, however, what with up to half the planks rotted out, people preferred to slog beside the road in bottomless mud. My experience has been what I describe. I lived through the roads around Fredericksburg. Mules mired and drowned there.

What I am saying also ought to evince what the Confederates knew from the start: Virginia favored defense as if created for it. What superb obstacles to invasion, those deep, abrupt ravines with streams crashing at the bottoms! Behind them the ranges of hills and their slopes, what positions for defense and created by nature! And forests to cover and keep secret the defenders' every move! And not only the treacherous streams that might betray the attackers' crossing, but also the horrid roads that might waylay troop passage, to render any consequential operation inconsequential from December to March!

One feature might seem to favor invaders. In all the South the dents open Virginia's coasts the widest to ships and the most vulnerable to attackers thus landed. Do not overestimate what at first looks perilous. Troops might get ashore but will they go anywhere then? Swamps border the coast, easy to seal, hard to pierce, the worst blocks to logistics. The names—Dismal Swamp, Alligator Swamp, Dragon Swamp, Rattlesnake Swamp—suggest little charm. Therefore the Union, for all the preparation and the equipment and the effort, penetrated nowhere and captured nothing by coastal invasion, unless you argue for New Orleans, a doubtful exception.

A secondary theater, adjacent to the eastern, enjoyed relations with it. So we must include the Shenandoah Valley, famous for fertility, beauty, and good roads. The Opequon and the Shenandoah flow through it perpendicular to one another, picturesque streams lined with willow, hickory, magnolia, and walnut. The valley parallels the Alleghenies and runs among the eastern ridges southwest-northeast. North, it joins the Potomac between Harpers Ferry and Williamsport. True, the Blue Ridge separated the valley from the seat of the war. But many "gaps" communicate through what would otherwise be impassable. They can be seen on the map. Notice them; they witnessed savage fighting.

2

What of communications in the state so invasion-proof in every other way? What, for instance, could the invader expect of the railroads, so important to war in the Confederacy? They played their role from the start, before I arrived. Often and doggedly the belligerents fought for them. Yet every line in Virginia deserves its place on the map, each offering one track to carry traffic in both directions. They often ran on bridges, brave affairs of wood, twenty to thirty feet above swamps for miles and therefore easily destroyed. Confederate rolling stock, in sad shape, had all it could do to meet the simplest demand.

These would be the lines for an invader: the Orange and Alexandria from Washington to Lynchburg and west; Aquia Creek and Richmond, of concern to the Union after 1864; and the Richmond and Gordonsville, connecting the two, an important link also for Confederate interior moves. Two spurs of the Orange and Alexandria, from Charlottesville and from Manassas Junction, served the Shenandoah but ended with no important connections. The one joining Lynchburg with Vicksburg on the Mississippi, via Knoxville, Chattanooga, and Jackson, meant the most for strategy, linking East with West. Virginia's lifelines connected her to the lower South: the Richmond and Columbia and the Richmond, Weldon and Wilmington. The RWW also went to Charleston. I mention others as they figure into this network and my discussion.

Railroads counted in both sides' strategies. What I say later will show the role. Let me say now that the South's repair shops barely sufficed at a pinch to fix and maintain weapons and materiel. The railroads' crippled cars might squeak and rattle at every joint; the shops could do practically nothing for them. The South must run its railroads ramshackle. By law therefore trains must not exceed twelve and a half miles an hour, to prevent accidents. Still, along the causeway between Richmond and Hamilton Crossing on the Rappahannock, derailments seemed to sow the fields right and left with cars broadcast. Some Northern shops, devoted to repairing and servicing trains, did nothing but keep them running; ergo Northern railroads in superb condition. Southern worsened until, by 1864, the troops on the way from Richmond to Petersburg followed as well as rode, walked, and kept up and took their turns riding. Yet the value of any train to the Confederacy cannot be overestimated. Trains delivered provisions until the final moments. Therefore the Confederacy spared nothing to rebuild tracks as fast as the enemy destroyed them.

What of the telegraph to strategy? I refer to the manual, though the high command would want the electric brought to it whatever the cost as soon as possible. The manual linked corps to corps. Indeed, every independent command had its signal outfit and their stations, an officer in charge and responsible to headquarters. In a long war the need to learn produced a remarkable system—even ways to communicate during campaigns. Lines and stations of the manual not only joined corps to corps but high command to outpost. Flags sent word by square meters of cloth, red to strike the sky, white for dark backgrounds. Semaphorists of this reliable service took them up the hill and the tree and waved so fast that dispatches took but minutes.

Codes must be changed often. A Confederate officer studied the Union's before Chancellorsville and broke it. Fourteen days ahead of the battle the enemy's intent could be read in his messages. The incident illustrates not only the need for secrecy, but also how rigorously the Confederates enforced it. I had just arrived to visit Lee in camp at Fredericksburg. Wanting to learn what I could of army and terrain, I rode out with staff reconnaissance. An officer in the ecstacy of discovery spilled the beans as we passed. He blabbed that he had broken the enemy's code. In the evening I congratulated Lee. Yes, I had heard the important news. His face darkened. Later my friend Von Borcke, the chief of staff, told me what happened next. Lee dismissed the chatterbox.

## 3

As to provisions, devastation made it impossible to get perishables from land raped by combat. So both sides ate what they could store and move. After the spring of 1863, Southern forces subsisted on little else than bacon, flour, and cornbread. The troops carried iron pans and baked the flour into bread, tasty enough. The North took full measure in victualing: fresh meat and canned vegetables and beverages, abundant, sumptuous. Jackson frequently told Lee: "I point out the herds grazing behind enemy lines. My men break through then, to get those treats."

Of course the need to carry food slowed and encumbered an army, crimped its style, and restricted campaigns to predetermined locales. The war from beginning to end reflects this proscription. Moreso indeed for Union forces, barely able to go a day or two from the warehouse. Confederates would load their simple fare on wagons; they had nothing else nor any way to bring along anything else. Little to move meant easier movement, so they turned poverty into luxury: few stores into many graces of movement. Lee, however, hated restraint. He groused even about that curb. He said to me: "Old Caesar always has the right words. He hits the nail on the head when he calls baggage trains *impedimenta*!" Packaged and preserved foods, and the elaborate means of moving and complicated ways of issuing them: thus feeding the men adds up to hindrance, obstruction, impedimenta indeed. Dependence on stores especially checked Grant. It kept him to the railroads or along the coast, flawing his otherwise apt campaigns in 1864. Only in regions spared devastation could his or any army free itself of impedimenta and eat by requisition; that is, live off the land by raids. Lee did it in 1862 and 1863, as soon as he carried the offensive across the Potomac. Sherman, too, marching through a Georgia still lush and spared until then the rape of war.

Nowadays it approaches blasphemy to speak of a "base of operations."[2] I defy the taboo, I blaspheme often in these pages. Remember that the term realizes its meaning in provisions and their supply in the American war. Sometimes, too, outlining strategy, I avoid repetition and do not stress enough the trammels of logistics. Please bear them in mind. They mattered.

## 4

I have been considering how generals dealt with logistics. It has brought me to the crux of strategy: generalship. Top commanders by and large lacked a plan early in the war. They vacillated. Brave ones commanded for the South. Even they shilly-shallied. Wanting to keep the enemy from crossing the border, for example, they tried with 60,000 men to defend its every foot, from the Kanawha River in West Virginia, across Kentucky, and as far as Missouri, some 750 miles. The South would have hurt itself had the North seized the opportunity thus offered. The North scattered the advantage, frittered away its strength, and spared the South a self-induced harm. My colleagues reading these pages will get the idea from the gross figures. How disorganized the leadership must have seemed in the West and especially in the early years! I call the early leadership the vagabond style.

Don't worry about localities; don't fight at the fringes. Break the enemy at the center; nullify his vital powers. So the principles of strategy dictate. At that time and place in the American war, though, the manner of warfare seemed ill-considered and ill-advised, a contradiction and a confounding of the principles. Reason enough that I explain the vagabond style of leading armies.

To begin with, war burst with little warning upon a vast region of few people. Along the borders, friend and foe intermingled nonetheless, indeed jostled one another, cheek by jowl. Families divided into factions; everybody reached for the gun; hostility sprang up everywhere; plenty of clashes must follow at random. These in time grew into engagements. Still the sides lacked organization—neither had a plan to marshal the fighting and coordinate the war as well as align the like-minded who implemented it—and both wanted leaders they could rally around and follow. Furthermore (I am still accounting for the vagabond style) organization and discipline produce and sustain an army unit by unit. Practice in directing masses of troops meanwhile creates generals. The Americans like any young nation therefore had to learn to fight and then to organize and direct the fighting. Finally the central government could not—it could not for sure all of a sudden—muster the human and material abundance of a nation and steer it to the one, the fatal end. The census provided data on population but not estimates of military strength to a government lacking the rigor to gather its strength. The army must be allowed to create itself before a strong hand brought order to impulse and concentration to confusion and aim to the struggle. Thus the stages; they must be passed; development would follow, stage by stage.

Yet, past them and having won the West, the Union did not unify its purposes in the West, ever. Rather than group at a point and break through, the Union tested the defense and tried to break through at too many points. Sherman at last did what ought to have been done. Under Grant the supreme commander, Sherman brought three armies together in the Battle of Atlanta, united, solidified, and sharpened for a *coup de grâce*.

## 5

In the East the Union contrived its objectives sooner and embroidered the grand design faster. A chaos of schemes had been hatched in part for adventure. Soon after Bull Run, however, clarity emerged and strategy took shape. By the spring of 1862 the armies had at least been arrayed and could face off organized. We want the capital of the rebellion, the Union said. Seize Richmond.

McClellan, with Richmond uppermost, began by laying before the president a plan for the assault. It took account of the situation and accorded with the circumstances. Indeed, Grant by energetic leadership would implement it and win at last. McClellan calculated early the cost of a drive straight to Richmond. Rivers threw up barriers; floods could make movement doubtful. Everything else about the terrain so favored defense as to add hurdles to barriers and further obstruct an offense. In dense forests you could neither see far nor find conditions favorable to holding a position if you did take it. The Confederates knew the way in those thickets. By now they had organized, too, and could square off against the invader. Inhabitants along the overland route would in their patriotism guide the Southern army as well as inform it of every hostile movement. McClellan, drafting his plan, appreciated the difficulties of approach overland. He would go by sea.

The navy, the service that had earned the North's esteem, by McClellan's plan would empower him to advance on Richmond via the James and the York. Naval elements not thus engaged would nonetheless cooperate, moving on Southern coastal places open to the sea, forcing the South to fragment its forces and weaken its resistance. The navy would also seize advantages thereby prepared. So a firewall would grow around the South and the fires of rebellion be banked and at last extinguished at the hearth. The Union called it the Anaconda Plan after the snake that crushes its prey in its coils. Would the plan yield victory? Yes—but not yet.

McClellan assumed the South would no more than put up a simple defense around their capital. He erred. He drew up a plan based on that error. He mounted a campaign to implement the plan. It failed. The Confederates' main force pinned down his army in terrain unfriendly to it along the Chickahominy, while the energetic Jackson led a large detachment to batter units scattered here and there and commanded by men subsidiary to McClellan but aspiring and overzealous. McClellan thus lost the troops under them, besides the losses to casualty and sickness. The Confederates unified and fighting, and how they fought, forced him to withdraw his weakened forces. A smaller enemy won by bold offensive against Union troops inefficient, undisciplined, and short on excellence.

The Union in its vanity did not blame those defects but censured his plan and how he coordinated its elements. Hence McClellan's loss of command and the shelving of the plan. It had been well laid—it remained to be followed to victory when Grant took it off the shelf and made it work. McClellan's other successors, Grant's predecessors, misinterpreted what happened to McClellan and tried the direct march on Richmond. Pope, Burnside, Hooker, Meade: each varied it, each failed. We shall consider the variations later. Grant too tried the direct march. Knock after hard knock taught him otherwise. When at last he subdued the remainder of the Confederate army, he operated as McClellan's plan prescribed: from a base on the James.

In the West the Union wanted first and foremost to gain the river and then extend the advantage into the valleys. The plan invited cooperation with the navy and its guns on that network. Brilliance put the Union in charge of the river from source to mouth by July of 1863. The success left the South divided and the North united to extend over Tennessee, whence to tighten around the rest of the South. Sherman won at Atlanta, broke through the defense, and marched through the heartland. Resistance ended. Rebellion had been extinguished.

## 6

We turn now to the Southern school of strategy for lessons. Though the subject tempt us, however, we shall not consider broad, ever-arguable theories. Consequences of Southern strategy pose theoretical questions and promise contributions to military science, true, but let my esteemed and learned colleagues search out, study, and answer them. The South fought like a master of strategic defense, true again. Our purpose being a basic, succinct account of the war as a whole, I repeat that we shall limit ourselves to how the South defended itself and what we ought to learn from the defense. Our brief but comprehensive concern promises to be nonetheless interesting and valuable to Prussian officers—if for no other reason than that Lee's campaigns resembled those of our Frederick the Great in 1756 and 1757.

Out of habit we may exaggerate the importance and value of the offensive. We regard strategic offense as the highlight of warfare, the mother of military science. Recall, however, that the strategic offensive requires by definition a near-parity of forces. Remember, too, that any offensive capable of destroying an enemy must be called strategic. Lee's thrusts north into enemy territory in 1862 and 1863 cannot be called strategic offensives; neither numbers nor goals meet the definition. The goals neither could nor should have included a sure and quick destruction of the enemy's power to make war. Lee lacked what his nation lacked but what that feat demanded: troops, troops, and troops. Like our great king's, his numbered nowhere near the enemy's. Rather, certain states being thought secessionist in spirit, his thrusts would agitate them in behalf of secession. Southern storehouses empty, furthermore, the thrusts would seize provisions to fill them. Moreover, as a Northern city would be a pawn in Southern hands to gain peace by negotiation, so the thrusts would capture one. They must be called not strategic offensives, then, but tactical maneuvers on a large scale. Of course, with this and every war, the obvious must be kept in view (it should go without our saying): a goal may be so to exhaust the enemy that he wants peace.

## 7

The Southern commander knew like few others how to use the keystone of strategic defense, interior lines. He managed strategic defense by getting the most from them. In our discussion of Southern strategy therefore we stress them. These shorter distances within a field differ from the longer ones of communication that link larger areas and serve other purposes, and from the formations arranged abreast of the enemy to do battle. I mean by the term the lines drawn when as many troops as can be spared at one point move in friendly territory to another where needed. Fresh, vigorous, assertive action can begin then and there, to continue until that element of the enemy can fight no more. Successful defenders then diminish themselves at that place and those that can be spared go via the interior line whence they can serve as reinforcements in the strategic scheme. Troops take their turn wherever and whenever needed, in a process lasting years. Thus the Confederates used the interior-line strategy—reinforcement, here larger, there smaller, as needed—and doubled their strength.

We need not look long for an example; the first days provide one. Others follow. Indeed, we can write the strategic history of the war in a chronology of the interior line. In July 1861, in the Shenandoah, all but four regiments of General Robert Patterson's Union volunteers disbanded for no good reason. (I return to the incident in another chapter.) Patterson's opponent, General Joseph E. Johnston, seeing the reduced numbers, on July 17 put the unneeded part of his army on a train to Bull Run. There, maneuvering by the sound of cannon, they happened upon the Union's right wing and panicked them into the retreat that made Bull Run famous.

The Peninsular campaign took place in Virginia: in the Shenandoah as well as on the York-James peninsula itself. McClellan with 100,000 men gestured from the peninsula toward Richmond. (I include the incident in my chapter summing up the war.) Opposite him and firm in defense stood Johnston and about 50,000. He remained so from mid-April to the end of May, using the time to dig in behind the swamps of the Chickahominy and to concentrate units scattered far and wide. Meanwhile, McClellan's right had failed to secure itself. Johnston, ready at last on May 31, attacked there.[3]

In the Shenandoah, Jackson and a 15,000–man division opposed a force three times the size. General Nathaniel P. Banks blocked the way to Harpers Ferry with a force larger than his. General John C. Frémont commanded two divisions in the western Alleghenies around Franklin. A little later, General Franz Sigel's division, part of McDowell's corps, moved into the Blue Ridge near Manassas Gap. Could Jackson succeed with an enemy in every direction, then, other than by determined, assertive interior lines? He did not have a capital at his back and the mission to defend it; he could move at will. Move he must, and move he did, and with interior lines, despite the small field. He went on the offensive south of Winchester against the enemies around him.

He attacked his most dangerous opponent, Banks, near Kernstown in March 1862. The battle, hard fought on either side, brought victory to neither. Jackson had a superb but small cavalry, only one hundred men. Even as Banks withdrew, Jackson detached a few, left them with the capable General Turner Ashby, and stole up the valley one hundred miles. There Frémont had crossed the mountains to cut off Jackson's retreat. Jackson struck—he had not been expected—and defeated General Robert H. Milroy's and General Robert C. Schenck's brigades on May 9 and pursued them over the mountains to Franklin.

He could not linger over the fruits of victory; he had heard of an enemy weakness. General James Shields's division had been detached from Banks's corps and sent after Confederates that turned out to be a figment of the imagination. It gave Jackson a chance; he must seize it. He marched another fifteen miles north past Luray to Front Royal. Colonel John R. Kenly held the pass through the Blue Ridge. But on May 23, Jackson scattered Kenly's troops and cut communication between McDowell and Banks. The bold Jackson could hit Banks without danger from McDowell now. Jackson and Banks collided near Winchester. This time Jackson had the larger force. Banks, routed on May 25, retreated in confusion. Jackson, in pursuit, scattered the last of Banks's forces on May 26, and chased them across the Potomac.

Jackson could give his weary troops no rest even yet. A new task called the Stonewall division south. Shields had been detached from Banks; now the Union leadership sent Shields to join Frémont, behind Jackson, to block Jackson's return south.[4] Jackson must counter. Leaving a detachment, he set out on a forced march from the Potomac to Harrisonburg. The rest of the army called his division "foot cavalry"; the 180 miles would not seem odd to men who could hoof it with and without horses. Only 13 or 14 miles lay between enemy forces at Harrisonburg and those at Port Republic. He arrived on June 8 at the last moment to put his in the sliver separating them.

He took on Frémont first—straight ahead on June 8 at Cross Keys, near and to the southwest of Harrisonburg—and beat him. On June 9 he battered Shields and set him to flight, chased him into June 10 and north as far as he could. On both days, Jackson's division had to deal with the pair, addressing at a given moment the one out front while the other's forays from another direction, however scattered and sporadic, must also be repelled. Fremont during this Jackson victory chose to withdraw across the Alleghenies to his old position.

Jackson's daring interior lines had at a stroke cleaned out the valley and put the enemy to flight. It also sent terror as far as Washington. It scared Lincoln into withdrawing McDowell's corps and General Louis Blenker's division from McClellan, adding 30,000 to the 50,000 defending the capital. Jackson's brave band thus tied up over 80,000 and hampered nearly as many, putting them on the defensive before Richmond. But interior lines had yet to be used with brilliance.

No sooner did Jackson rock and stagger his opposition than, leaving cavalry and an observation detachment, he moved on foot and by rail as fast as possible the two hundred miles to the seat of war before Richmond. The moment had arrived for the brilliant use of interior lines. Jackson used them with brilliance. Meeting detached allies along the way, he added their numbers to his. He went fast enough to reach Gaines Mill by June 27. He joined the battle at the apt moment—no wonder Lee wanted him there then—hitting McClellan's right wing and bringing the South a triumph.

We can understand McClellan's astonishment and anger. To keep Jackson in the Shenandoah, McClellan weakened himself by some 50,000. Now, accordingly, he believed Jackson far, far off. Suddenly Jackson confronts McClellan and undoes in a flash everything McClellan labored at.

Advancing upon Richmond, McClellan sent reconnaissance ahead. Defeated, he inched back and watched those detachments pushed after him. The Confederates, to meet him before Richmond, had weakened their positions along the coast—bled them to embarrassment. Now, understanding strategic defense and interior lines, the Confederates returned forces to the coasts at the rate that he retreated.

General John Pope's advance on Richmond from the north came next. Pope massed forces. Learning of it, Confederates in Virginia divided again, this time into operations along the Rapidan and near the James. A glance at a map shows the lines of communication that must result. The Richmond and Gordonsville Railroad linked the Confederate fronts. The Union would go by sea from Washington to the James, thence overland to the Rapidan. It follows that the Confederates had it over the Federals in communications. Lee, heading the Army of Northern Virginia since the Battle of Seven Pines, exploited the advantage—need we even say so? Unfortunately, before withdrawing to the new theater, he could not teach McClellan a last lesson.[5]

He had found McClellan—the general of engineers, the master of encampment and fortification—too well entrenched. Lee must be satisfied to leave a weak detachment to watch him. Lee knew others could be sent south by rail to parry a McClellan offensive. McClellan tried a weak, scattered one, August 5, 1862.[6] As for Pope, we saw earlier in the chapter what happened to him. The South mustered around him the power to smash him with a vengeance.

Next the Confederacy advanced into Maryland, then reached into Kentucky and Missouri in October.[7] Did the Confederacy use pauses to send troops from the East to aid operations in the West? And return them to fight in the East? Neither reports nor other sources confirm it. But, though people watched their tongues when discussing such intelligence as troop strength, enough came to me that I take it for a fact that the South sent troops to and fro and by interior lines. We know the South understood and had used the lines enough to extend them that far by then. Subsequent events support my conclusion. Recall for example that at the end of October, as operations revived in Virginia, so they slackened in the West.

Burnside in December led the next Union attempt on Richmond. The Confederacy put the quietus to it at Fredericksburg. In January the Union tried a "mud campaign" and mired in failure.[8] Lee had fought to mastery; he controlled the theater. In a post that commanded troops and respect, he ran the show from south of the Rappahannock until the spring of 1863. On the peninsula, General John A. Dix gestured at Richmond. Union forces also moved into North Carolina, aiming for the Weldon Railroad. Lee, using battle-earned superiority, detached generals: Daniel Hill and his small corps to North Carolina, and James Longstreet and his three divisions against Dix. Lee kept only about 30,000 men and stayed inactive meanwhile. Only his cavalry kept harassing the enemy and informing him of enemy moves.

He therefore knew in time to counter them. He knew of Hooker's campaign before it started, thanks to the cavalry and his network of scouts. So he could counter Hooker. Many authors, believing falsehoods uttered by General Alfred Pleasonton in the committee proceedings, have assumed Lee knew nothing of Hooker's intentions until April 30. I need only speak from my experience on Stuart's staff. Stuart received dispatches as early as April 28. At 5:00 A.M. the next morning—so soon, suddenly—he ordered us into the saddle. My fellow officers of the staff whispered to me: "A 'big battle' brewing." We rode with General Fitzhugh Lee's brigade straight to Kelly's Ford to watch Hooker's crossing of the Rapidan. Did Hooker know we watched? Did he even know Stuart's whereabouts? We got our answers when the officer looked at us astounded to see us then and there.[9]

8

We observe in the Hooker episode to our profit the mastery of strategic defense. We learn also by negative examples: a well-laid plan gone awry because the planner did not rightly estimate the enemy; then not being as flexible as elastic so as to reshape to any irregularity and adapt to each surprise by a bold enemy; then to suffer such overelaboration that a nudge shatters a system rigid in details; and finally the consequences to drawing a plan in red but carrying it out in yellow.[10]

Hooker moved with 150,000 against Lee's 50,000. Lee held the solid position south of Fredericksburg. Hooker's plan—I mention it in my overview of the war—would have General John Sedgwick feint with 40,000 and occupy Lee at his front. Hooker meanwhile would take the rest up the Rappahannock and cross the Rapidan to strike Lee in the flank and rear. Lee, according to the plan, must either leave or be cut off from Richmond and caught in a crossfire. Hooker laid the plan with deliberation in consultation with informed commanders, knowing the terrain and, so important here, aware of the enemy's strength and apprised of his situation. Plan excellent, advice savvy, he could have succeeded with the superiority of numbers at his disposal. Only what had served him in lesser commands failed him here. He lacked what he once possessed and needed most now, what the general needs above all after laying the plan; the gumption to carry it out. He must follow thunder with lightning, and lightning with a strike. Planning and execution, together they distinguish the commander from the planner and the steward. Plan, aye, but then deal with weather, friction, strife, and faulty coordination on the march; overcome what rises to block you; devise measures to counter offensives and counterattacks; and make the most of enemy weaknesses! In other words in the last analysis the general must have initiative, vigor, and *drive*.

Hooker began right. But instead of loosing his host on the enemy, he paused, that his host be properly drawn up (thus the planner and the steward, not the commander). Attacked, he did not use superior power. He shilly-shallied. Even on May 2, 1863, well into the tragedy of the Wilderness at Chancellorsville, when nothing but a counterattack could free him, he hesitated and—cowed, browbeaten—let the enemy seize the initiative. Shackled, he did not exert the strength to break out during the following days and the crucial moments. He withdrew to the far shore of the Rappahannock and guaranteed his own defeat. Robust of design, sickly of execution, craven after boldness, plan and maker invite comparison with the timid little farmer on the noble, fiery thoroughbred. Away he goes, unfit to ride and out of control. One leap throws him.

How different with Lee! Circumstances favor Hooker. Lee disregards them. A mediocre commander—and if, like Lee, undermanned to face this burly foe—would fix himself on defense, curbing movement. Lee leaves Hill in place at Fredericksburg and attacks. Destiny may offer a chance. He watches for it. Seeing the enemy vacillate, he knows the enemy lets himself be cowed. He hits him in the flank and rear, and rattles him. He directs the troops with the vigor of the maestro in concert. The day, May 2, ends with him on the upbeat and pressing to the next measure.

After that effort, did his troops retain the wherewithal to storm the enemy's second line? I cannot lie. They could have. Why didn't they, then? (Questions come pell-mell now.) I could only be disappointed at success not carried to the limit. The troops jumbled and confused? Yes. They had attained a superb objective already? Yes. But we must ask the right question. What of first prize, the final goal? A wide river cramped the enemy and impeded his withdrawal, making the prize grander and easier. Lee of all generals ought not have been guilty of hesitation and second thoughts, certainly not when he had succeeded so far and conditions favored more success. He ought to have exploited success as Jackson usually exploited it—and to the last breath. Lee spent a day of doubt, pausing after victory, hesitant about Hooker, pausing after victory to indulge second thoughts.

On May 4, Lee took charge, master of the situation again, having mastered himself. Hill had forced Sedgwick out of position near Fredericksburg. Leaving Hooker dazed, Lee, with forces reduced to a pair of divisions, led them to a surprise on Sedgwick and defeated him and threw him with losses back across the Potomac at Banks Ford.

## 9

Lee tried next to multiply the gains of Chancellorsville. The Army of Northern Virginia must be reinforced; every man who could be spared, withdrawn from south and west; and interior lines tested, to increase the army to about 75,000 for this offensive. He would do his best in the face of an enemy north of the Rappahannock and twice as strong—he would try no less than to capture a few cities and compel peace on terms agreeable to the South. What a dare! His government defied the risk and put all behind every hope of success, even stripping the capital and leaving itself almost naked in the dare's behalf.

I took part in the Gettysburg campaign. Excuse me, colleagues, if I spend a while on it.

Lee began with a cavalry gesture while keeping back the rear guard under Hill to hold the enemy on the Rappahannock. Then, behind General Richard S. Ewell's corps, Lee hurried around the enemy's right, unnoticed. He proceeded in secrecy; the Federals knew nothing for sure until Confederates stumbled on Milroy at Winchester. In Milroy's shriek the Federals heard the first sign of Lee's intent.

Lee's army became three corps when Hill joined after taking Winchester. North they went arrayed for mutual reinforcement: one behind the other at intervals of one or two days' march. In one or a couple of days, two corps—and in another day or so the army of three—had been disposed to prevent being split. His right toward the enemy, Lee deployed cavalry toward the enemy there. The columns to that side, moving thus under the cavalry's eye and protection, could spring to reinforce the cavalry against attack. A map will show

how, in that formation, the Confederates swung in a half-circle below and west of Washington. At the end, the headwaters of the Monocacy, they met the enemy, closed ranks, and fought the three battles at Gettysburg. Outcome: indecisive. The Confederates must fall back along their lines of communication south, lest the Confederacy be subject to a stroke of fate. Lee's forces withdrew to the old site on the Rappahannock.

I think Richmond, to strengthen Lee, had so weakened itself that the Union meanwhile could have taken it with a strong detachment. Numerically superior, the Union could have spared the detachment. Taking Richmond, it would hold the Confederacy's arsenals and the seat of civilian and military administrations. Lee, fearing the very thing, paused after Winchester while his rear still covered the Rappahannock. He breathed easier when the enemy turned north to describe between him and Washington a smaller while he described a larger half-circle. Hooker sensed the potential of a thrust south and laid the plan before the cabinet in Washington. Lincoln returned a characteristic reply in autograph.[11]

> I have but one idea which I think worth suggesting to you, and that is, in case you find Lee coming to the North of the Rappahannock, I would by no means cross to the South of it. If he should leave a rear force at Fredericksburg, tempting you to fall upon it, it would fight in entrenchments and have you at a disadvantage, and so, man for man, worst you at that point, while his main force would in some way be getting an advantage of you Northward. In one word, I would not take any risk of being entangled up on the river like an ox jumped half over a fence and liable to be torn by dogs front and rear without a fair chance to gore one way or kick the other.

That letter and the documents of the time show that the cabinet directed the army's every strategic move and that the directives exceeded Austria in narrow-mindedness.[12] Such Byzantine sovereignty cannot render a prompt, decisive, skillful decision. Action then comes after pseudointelligent deliberation cankers what it directs. Action so directed resembles the fruit out of season that drops from the branch, rotten. The war could turn better for the Union only when Grant freed himself of those shackles.

Let me discuss some of the flaws consequent on such leadership in the days after Lee began the march to Gettysburg. General Pleasonton led the large Federal cavalry then. An example will show how it failed the Union. By June 10, 1863, Ewell had departed for the Shenandoah; yet on the twelfth, Hooker (left ignorant by the cavalry) telegraphed Dix: "All of Lee's army, so far as I know, is extended along the immediate banks of the Rappahannock, from Hamilton's Crossing to Culpepper." This when Lee had gone on June 7 and 8! Example abound. Pleasonton's statements make an eyewitness laugh.[13] He said the campaign began when he attacked the Rebel cavalry on June 9, 1863 at Beverly Ford (Brandy Station to the Confederates). He also said the Rebels suffered defeat there. I saw the opposite. He claimed further that he learned by this battle that Lee would invade Pennsylvania; yet his letter to Butterfield on June 12, 1863 shows otherwise. Hooker, learning thus of Lee's designs on Pennsylvania (according to Pleasonton), moved his army to Maryland. Finally, Pleasonton claims he engaged the Rebels on June 17, 19 and 21, at Aldie, Middleburg, and Upperville, and with such success that Lee had to abandon his plan to cross the Potomac at Poolsville and go to Chambersburg via Williamsport and Hagerstown. My diary for June 20 says, "We've gotten word that Ewell has crossed the Potomac at Sharpsburg, and Rodes at Williamsport."[14]

What do those three cavalry skirmishes of June 17 to 21 mean, really? They prove the harmony in the Confederate formation, how efficiently, how effectively the parts worked together on the march north. Let me illustrate. At that time a letter from Stuart told me Von Borcke had been wounded.[15] Come here, Stuart said. I arrived at Upperville at *the* moment. Stuart could no longer hold against superior numbers. Struggling with the difficulty of his wound (a bad one), I got Von Borcke to a safe place. Stuart's cavalry had been pushed back only to the top of the pass near Paris, however, when columns of the formation approached. Advance elements appeared. They secured the pass and recovered the lost position, at the same time reversing what might have been the end of Von Borcke. Of course none of this action deflected the main army; it did not swing our way but continued north undeterred.

What did Gettysburg mean to the war as a whole? The South had strained beyond the limit. The Pennsylvania campaign failed with ugly consequences as it absorbed the last of the nation's wherewithal. The West had been bled of troops; hence the loss of Vicksburg; thus not a link between the South's East and far West. General Braxton Bragg lost Missionary Ridge for the same reason. Lee had saved troops on his strategic defensive; he returned them west in the fall of 1863; but so what, then? What good that troops from Virginia helped recoup Bragg's defeat with victory at Chickamauga?[16] Who cares that because the South reinforced along the Mississippi, every Union effort failed there henceforth? The fact remains, the South had enjoyed defensive superiority by interior lines. It lost it by defeat in its one thrust north. The South could do nothing but try to defend itself on every front then. A strategic defense depends upon one essential, and the theory makes it plain, that at least part of the forces be capable of offense.

## 10

Lee could not conduct such an offensive—he had lost the capability of strategic defense—but his example remains instructive. Therefore we cannot ignore, we must study, his 1864 strategy of tactical defense. Indeed, in the campaigns against Grant, from the Wilderness to Petersburg, we watch the masters at the final, the championship, match. If I may extend the figure, Grant can be said to prefer the rook. By never swerving from the rook's way, he intended to dominate the game and end it. He had nearly 195,000 troops and deployed for their straightforward power of numbers against Lee's mere quarter the amount. Lacking the power of numbers, Lee took the more elegant and graceful way of the knight and the bishop.[17]

Grant moved first. By virtue of an opening in the line, he intended to flank Lee on the left; then with a push to Bowling Green, establish a supply route for himself and cut Lee off from Richmond. Lee countered. He met Grant on the other side of the Rapidan in the famous Wilderness. Lee went on the offensive because he wanted the initiative. He wanted it as soon as possible. He failed because he lost Longstreet to a Minié ball in the neck. Lee's move gone wrong, Grant sought advantage from it. He reached for Spotsylvania Courthouse around Lee's right. Lee, anticipating, made the crucial move against Grant's forward units: wheeling with snap: blocking by covering approaches, yet staying on defense, safe, secure. Grant, reinforced, gestured as if to turn Lee's right, go for Richmond, and declare checkmate. Lee countered with part of his army against Grant's wing opposite. Grant mistook the part for the whole. Lee meanwhile, by forced march, inserted his army between Grant and the threatened capital. He had left Grant marching down the road south. Grant did not discover his mistake until he reached the Pamunkey River.

Hastening to rectify, he crossed to Hanovertown, trying again to flank: his fourth such move, as adroit as the others, and an equal flop. Lee blocked him, ready to fight. Blocked, outsmarted, supply line cut, Grant now if ever knew what it meant to face the master of the art of workaday warfare. He went to White House, from there to try the third time to drive by force through that mobile ring around Richmond. He drove at Cold Harbor and must withdraw to lick his wounds. He gave up trying to win with the drive by force.

## 11

The war ends there, for our purposes. Grant and the North had lost that campaign and 60,000 men. Grant found himself where McClellan began against Richmond in 1862. The war of pick and shovel started at Petersburg in June 1864. It offers little to our study.[18] Others of the war's episodes prove interesting but for us not as instructive as what we have considered. The prime example, Lee's campaigns—arduous, pivotal, executed as masterpieces—featured the brilliant use of interior lines. Lee, with them, doubled his strength without adding numbers and prolonged the war for years without being reinforced. Grant could not end the war until he adopted and stuck to the means that would work: pin down the army that mattered; make too hard detachment from it; overpower it; squeeze it to exhaustion; and with five times its numbers, crush it. Sherman cut the artery and ended resistance by the South's last handful. Grant won, Sherman conducted a superb campaign, but Lee's work must go down in history in nothing less than golden letters.

# 2

# TACTICS

*Armies squared off—armies of an intelligent, practical nation—but armies as yet unschooled.*

Background—First Period: Development, 1861—Elaboration of Linear Tactics, 1862-1863—Perfection and Adoption of Defensive Tactics, 1864 to the End—Comparison with European, German in Particular.

**B**efore I describe tactics in this war, let me glance backward at the influence, upon warfare in Europe, of the first American war, the so-called War of the American Revolution.

At the end of the last century, the century of that war, Europeans still fought in lines. Frederick the Great perfected linear tactics and made them standard. Men of less genius than our great Prussian king perverted them, made them pedantic schematics of finicky, fussy evolutions and the formal, routine firing of salvos. These tactics, even as perverted, had value on the plain of Europe. In the American forest they hatched dinosaurs. The rebels did not know the British drill but did possess the common sense that has always distinguished people across the Atlantic. They had also got used to the peculiarities of the terrain as a place to fight in wars against the Indians. Opposing the rigid lines of the British, the Americans used flexible skirmisher tactics, fought a war of diffusion and dispersal, and swarmed among the enemy's helpless masses. However defective their military instruction, they drove off the enemy, chased all the king's men out of their country, and put it back together for themselves again.

This skirmisher warfare (the new, the American way), spread to the Continent in written accounts. A few Europeans knew of them then. They, the smug, highfalutin Europeans, scorned them as backwoods. Except the French. After their revolution they (like the Americans) had raw troops. The revolution's levies raised multitudes, but the rush of events left no time to train in linear, so they decided to profit from the innovative, the American tactics. The French taught them to their rabble in arms. Heavy, solid columns, with skirmishers out front for protection and attack, created their own center of gravity and gave the light, mobile skirmish lines stability and support. With those tactics the French emperor trampled half of Europe. Our own Prussian, pioneer in linear, learned in a few hours and after intrepid resistance: skirmisher tactics had superseded linear. Especially Prussia must suffer because it closed its eyes to the art of war elsewhere.

Did the Americans in their second revolution introduce tactics that would again be themselves revolutionary? No. Nor can it be said that we shut our eyes again to innovations abroad. Still the tactics considered in these pages, in particular the later developments, may provide every thoughtful soldier a wealth of ideas for collation as well as new material for anyone interested in tactics.

Why, then, the fascination of this war? America had proved itself imaginative, inventive, and ingenious—qualities that flourished, flowered, and bore fruit there. The world would therefore watch when two armies squared off there—the armies of an intelligent, practical nation, but armies as yet unschooled. To reach the goal, either side would try anything and everything. No rule, custom, practice, regulation, predetermination, or prejudice would curb the spirit. Ways new to North America would not be scorned. The queerest of weapons would be tested; let the enemy beware. From the chaos of attempts, experiments, trials, tests, thrusts, campaigns, and expeditions, infantry tactics developed clear and sharp. Both sides adopted them. They deserve our attention. Let us consider them in their several stages.

## 2

The first began with an outbreak so sudden that friend and foe had not yet separated but intermingled still, especially on the borders and in the Border States. A state of war existed but not the means to wage it. Frictions, brushes, clashes, skirmishes, fights—at most, engagements—occurred: enough to ignite and fan hatred, not enough for anything else. When the presidents called for volunteers, neither side had known war yet.

Thus, ignorance of the military prevailed in the ranks of the military, if they can be called military, if they can be called ranks. Haphazard clutches of warriors must have been amusing in their fantastic garb and bizarre equipment. Should they be called the soldiers of a modern nation? Or bandits with revolvers and bowie knives thrust into their sashes?

Behavior confirmed looks in the early action, the days of braggadocio and the interlude of romance. Ruffians roved in scattered, independent gangs, raiding, ambushing, and hatching the strategems that signify partisans. Reports tell of feats of arms in phrases as eccentric as the participants appeared and as disjointed as they acted.

Given time, Northern factories produced more and more, and better and better, and then excellent weapons. Union forces, armed to the teeth, tried tactics based on the firearm and the fire fight. They blasted distances and hours at enemies they thought hidden in the forest, in fact nonexistent. Artillery distinguished itself the same way at first on fictitious targets in the antipodes. Weapons had become a fetish. Right off indeed, in the First Bull Run, Union batteries opened up on Confederate positions at nearly a mile and three quarters!

On the Confederate side, inferior muskets and antique shotguns carried nearly as far and about as accurately as the Confederacy's smoothbored mortars—one-pounders, two-pounders, three-pounders—on jerry-built carriages. So the Confederates tried to close in and decide by cold steel.

Those extremes taught sides to seek the golden mean in tactics. Differences in weapons lessened, too, as the Confederates got better armed, mostly by capture. Contrasts remained, however. Confederate inferiority in arms cannot be missed throughout this, the first stage.

Command had to be as primitive as tactics when men on both sides understood little, therefore as much as most of the officers. Leadership could not grow nor discipline thrive in ignorance. Action usually succeeded to the extent the majority thought it worthwhile. Attempts to consolidate, with higher officers heading larger units, proved superficial, more in name than in fact, and never approaching the impossible, the niceties of organization. Popular opinion distrusted adjutants and staffs; they decorated the general with "feudal" pomp. He—all officers—therefore must get along unaided. Rather than concentrate on campaigns and operations, he must worry about nuts and bolts in the machinery that did not work anyhow. Still, by the eve of First Bull Run, the Union had shuffled into five divisions; the Confederacy, nine and a half brigades.

Our first period climaxed, burned itself out, in the heat of Bull Run. The denoument, though cooler, proved more important. After a few hours—I mention it in another chapter—the Union army panicked for no good reason, or so we hear, and fled in funk to Washington and beyond. Despondent, fainthearted, they degenerated from disorganization to chaos.

Not the Federals, however, but the Confederates amazed military circles elsewhere. Rampant round them ran the claim that the victors could not follow through. "They could have had the enemy's capital but stopped short. They could have had it maybe without drawing a sword."

I do not hold that view. To me, not what the Confederates did, but the amazement at it, constitutes the blunder. The Confederate army stalled for clear and cogent reasons. Rebel as well as Federal spirits less than stout had been shaken. Neither side wanted to persevere; both wanted to run. On the Southern side, too, the regiments had mingled, confused. Patrolling, sentry duty, other duties, and on and on, besides the unnecessary state of alert had played their parts. Unduly, devastatingly, undoubtedly the added stresses and strains tired inexperienced forces troubled with the normal stresses and strains of soldiering. True, they had not defeated the enemy; they had routed him. But could they have pursued him? They, victorious but no better than he in this wreckage, could not have pursued him a step. Pursue the fleeing foe, then? Even get an inkling of the size of victory? Contrary to the implications of those questions in the mouths of critics, the Confederates had done all they could. Men and officers of the Confederacy furthermore expected to have to fight again the next morning. They did not know their success until they talked to local people and read the newspapers and learned how fatigue, ignorance, and fear had weakened and blinded them. They had won but, left in such a state they could not know they had won, how could they follow through?

Bull Run taught them at least one lesson. Even the most spirited mobs cannot fight a war. The lack of intermediate levels of command had made itself felt, too, drove itself home. The general found it next to impossible to direct nine brigades himself. Southern commanders did not pursue the enemy; they set to *the* task, *the* problem: organization. Barely had the dust of combat settled but rearrangement began with the articulation into corps and divisions.

The North must deal with worse: defeat. The dreams of victory, the confident prophecies, the rash and precipitate rejoicing—all must be forgotten. The nation must be frank and ashamed, confess folly, and be contrite. The army had been devastated. To the Union's credit it must be said that the people retained strength to take stock and buck up and do so at once and not complain of treachery and betrayal. They had regarded everything as a business deal. Yankee traders out to win a war as they would make a dollar, the shrewd Yankee businessman, had seen the situation in terms of trade, a question of supply and demand, an application of horse sense. With defeat, feelings changed to duty and honor. The people examined conscience and sought redemption in the return to logic and discretion. Only one victim must expiate the communal sin. They sacrificed the old general. Famous and popular in the Mexican War, hailed by the press then, a hero for the time, Winfield Scott, though ill now, would serve his country this one last time. They subjected him to cheap criticism. He, the patriot, must take it on the chin. Bull Run had been in July; he took the hard knocks and retired in October.

The North thus suffered the loss of confidence. The opposite became a problem for the South. After victory at Bull Run the populace and the average citizen thought too well of their Confederacy and, pompous and conceited, developed disrespect for the Union. The attitude, developing even before Bull Run, cost there and then, and all the more in the next stage. Like every overweening opinion among backers, this one brought harm to the Cause. They would pay in bitter experiences. The general saw the evil but worked against it in vain. As late as the campaigns of 1863, Lee complained to me that this false confidence caused him distress.

Let me affirm here that I refer mainly to events and conditions in the East. Railroads did carry troops East to West and back, ever exchanging innovations. But the East innovated, the West imitated. The East initiated, the West followed by a step or two. Armies in Virginia also offer the best examples of organization. Besides, the Union merely won the West. In the East the Union sealed the Confederacy's fate.

I have said that the first stage climaxed at Bull Run. Let me note that if the battle opened eyes on their side of the Atlantic, it closed them on ours. Europe's military establishments lost interest in the war. After the lame performances by both sides at Bull Run, not one country's officer corps paid much attention to the war over there. With an air of superiority they shrugged and laid aside the reports and ignored a war that as it developed could have taught them many a lesson in more than one aspect of warfare.[1]

In America the battle opened the eyes of even the blindest booster of militia: from that moment on, everybody, North and South, wanted an army fit for military service, no more citizen soldiers and minutemen. Every effort went to making an army that if not out-and-out professional, could fight like professionals.

Would the nation not want something better than militia, given words like General Irvin McDowell's on Bull Run in the official report? Let these excerpts illustrate the troops' condition as well as my point, and show the odd role of troops known as the "three-months' volunteers." The first shows the state of McDowell's forces; the second, what it meant to lead men of short-term service.[2]

(1) "The ridge being held in this way, the retreating current passed slowly through Centreville to the rear. The enemy followed us from the ford as far as Cub Run, and owing to the road becoming blocked up at the crossing, caused us much damage there, for the artillery could not pass, and several pieces and caissons had to be abandoned. In the panic the horses hauling the caissons and ammunition were cut from their places by persons to escape with, and in this way much confusion was caused, the panic aggravated, and the road encumbered. Not only were pieces of artillery lost, but also many of the ambulances carrying the wounded.

"By sundown most of our men had gotten behind Centreville ridge, and it became a question whether we should or not endeavor to make a stand there. The condition of our artillery and its ammunition, and the want of food for the men, who had generally abandoned or thrown away all that had been issued the day before, and the utter disorganization and consequent demoralization of the mass of the Army, seemed to all who were near enough to be consulted—division and brigade commanders and staff—to admit of no alternative but to fall back; the more so as the position at Blackburn's Ford was then in the possession of the enemy.

"On sending the officers of the staff to the different camps, they found, as they reported to me, that our decision had been anticipated by the troops, most of those who had come in from the front being already on the road to the rear, the panic with which they came in still continuing and hurrying them along."

(2) "I could not, as I have said, move earlier or push on faster, nor could I delay. A large and best part, so considered, of my forces were three-months' volunteers, whose terms of service were about expiring, but who were sent forward as having long enough to serve for the purpose of the expedition.

"On the eve of the battle the Fourth Pennsylvania Regiment of Volunteers, and the battery of Volunteer Artillery of the Eighth New York Militia, whose term of service expired, insisted on their discharge. I wrote to regiment as pressing a request as I could pen, and the honorable Secretary of War, who was at the time on the ground, tried to induce the battery to remain at least five days, but in vain. They insisted on their discharge that night. It was granted; and the next morning, when the Army moved forward into battle, these troops moved to the rear to the sound of the enemy's cannon.

"In the next few days, day by day I should have lost ten thousand of the best armed, drilled, officered, and disciplined troops in the Army. In other words, every day which added to the strength of the enemy made us weaker."

Let me add a bizarre coincidence. The Union general Robert Patterson, opposing Johnston in the Shenandoah, lost an army the same way in one day, 20,000 men. Only four regiments remained to do as he ordered. The rest claimed that their states treated them shabbily with spoiled pork and often nothing but two or three crackers a day. Consequently he withdrew with the entire force. Johnston could send his to Bull Run. They decided the battle. When the sad news reached Patterson's men, the cowards hissed Patterson during a parade: they whose dastardly behavior partly caused the defeat. They raged at his leadership, accusing him of wavering because he had not ordered them forward against their will.

In the North the professionalization of the army, the labors of Hercules, fell to McClellan. Developing the Union forces, he devoted such perseverance and applied such skill as to secure forever the reputation of organizer. First he sought to infuse the officer corps with martial spirit. He established examining commissions to cull the useless. He got rid of over one hundred drones himself. To assemble and systematize, he created standard and typical units and regulated conditions and terms of service. He created divisions and charged them with functions lacking before. He originated food and munitions transport, as well as the rail service the medical corps needed. He developed staffs for commanders and specified their duties. Applying discipline and enforcing the regulations for the government of armies in the field, he brought obedience to the ranks and order to operations.[3] After seven months' work—he dared not tire—he could risk leaving camp and going south before the season's abundant rains ended.

The first maneuver, a test, turned out badly of course. Defects had not been removed nor malfunctions fixed; the mechanism had been asked to run in spite of them. The army spent eight days in disarray, did not find the enemy, let alone engage him, and returned to camp. It must be a campaign of short marches and long pauses. The army had to learn to march and to pause, and through them to conduct itself and manage tactics in the battles before Richmond. Indeed, in learning the preliminaries, the army showed itself learning battle itself. True, divisions panicked often enough. But marching to the James through the Chickahominy swamps and in battle every step, the inexperienced proved they could be superior.

Inner workings, then, must be set right, else the army not strike with effect. The South, as we have seen, had to learn the lesson, too. The South enjoyed capable directors (one advantage) and a president versed in military affairs (another advantage) to do the same. More success accrued to the South accordingly; and from their swarms of volunteers, a "crew" disciplined and ready to fight; and superiority to the larger enemy, David to Goliath, every time they met. Especially Jackson's shone forth by their distinction: model to the armies. Jackson in secret shaped them to march and fight. As early as the spring of 1862 in the Shenandoah, they marched and fought without interruption and onto military history's honor rolls.

## 3

Our discussion of early combat has brought us to the second stage and battle formations themselves. Much of what I know I learned in fourteen battles and skirmishes. What I say follows what I learned.

Americans tried the column for offense and gave it up because artillery poured murder on their columns. In their linear tactics on offense a dense formation of skirmishers arrayed themselves in their line about 120 to 150 paces before the initial line of infantrymen. Dense indeed: one skirmisher to every four men in the initial line! The ratio would be about one German battalion on line and one company in front as skirmishers. A second line followed about 150 paces back. A third stayed in columns about 250 back. Troops formed for battle, I must stress, before they entered onto the field; they had not the skill to form under fire. I mean that both first and second formed ahead of time, for even the second could not cope with the niceties of array in new terrain while being shot at.

The parade-ground layout, the textbook formation, shows a division arrayed in three brigades, or two regiments each, one behind the other, the first and second brigades 150 paces apart; second and third, 250: the first in a long, single rank; second, in a rank with supports and reinforcements to the rear; and third in two wedges; with reserves to their rear.

The typical brigade consisted of three or four regiments, usually paired abreast. In battle the brigade, thus disposed, became the tactical unit. In other words, lines for battle maintained individual integrity and interlinear relationships within the brigade. I would almost say they defined themselves by the brigade and pledged allegiance to it through thick and thin. Regiments and brigades in cooperation constituted a division. Deployed for attack, they looked like ghosts of days and ways of Frederick the Great. Forward they went by a command, not loud but passed in a murmur along them from captain to captain. Skirmishers' fire prepared the attack. The nearer to the enemy, the more faulty the lines and the more ragged the first until it crumbled and mixed with the skirmishers. Forward went this muddle leading the wavy rest. Finally the mass obtruded upon the point of attack. In a sustained, stubborn clash, even the third would join the melee. Meanwhile the usually weak reserve tried to be useful on the flanks, or stiffened places that faltered, or plugged holes. In sum it had been a division neatly drawn up. Now its units, anything but neat, vaguely coherent, resembled a swarm of skirmishers.

Confederates in this disarray would have had to use the last reserves, few at most, at the first crisis; only, encouraged and guided, the troops developed the tenacity, the doggedness to fight many a battle to the end in just such a swirling muddle. Of course, however brilliant the success, they could seldom follow through. Such a mass can be but badly led. Furthermore, anyone who knows war will confirm that after hours of ferocity the mass in battle can no longer manage a thrust, let alone force an advance. And follow through with reserves the Confederates never had? Confederates who won could not pursue; they must settle for having set the enemy back. Even Jackson, first-class tactician, could only now and then pursue to the last gasp.

With those words I have sketched a division in the attack. I need not mention again that on the battlefield it took every modification, and in each clash, configured itself anew. On the whole, however, I described the elements of the battles of 1862 and 1863.

To leadership in battle on wooded terrain the top general mattered the least. He could see nothing of the fight. Only in the rarest instances, as I have suggested, did he keep a reserve with him. Lee did not hesitate to speak to me about it, three days after Gettysburg. "You have to realize how things stand with us. Recognize that my orders then would do more harm than good. I rely on my division and brigade commanders. How terrible if I could not. I plan and work as hard as I can to bring the troops to the right place at the right time. I have done my duty then. The moment I order them forward, I put the battle and the fate of my army in the hands of God." At Chancellorsville, May 3, 1863, we—I at his side—watched the fighting reach crisis. We stood in heavy enough fire while an interesting episode played out before us. Even as excitement at the progress gripped the great captain, he spoke to me of education: what he wanted of it for his nation. The subject astounded me there, the more because this leader, as closemouthed as our revered Moltke, said little anytime anywhere.[4] Often he said only what he must to do his work.

In this period, striving to take and hold the initiative, he seized the tactical offensive. The purpose dictated attack even under what looked like unfavorable circumstances. On the third day of Chancellorsville, May 2, 50,000 Confederates, in a frontal attack, dislodged three times as many Federals from stout fortifications. Let not one example, Chancellorsville, suffice. Consider Jackson in the Shenandoah, as well as these: Gettysburg, the campaign against Pope, and the Seven Days before Richmond—Confederate offensives. Only at Sharpsburg and Fredericksburg did the Union attack. Otherwise the Union gestured at strategic offense but kept to tactical defense.

Tactical defense. Tactical offense. Let us go back to the start, review events, and thereby elaborate and clarify those terms. The American loved defense behind the lightest of fortifications. Likewise he valued positions easy to defend well. Yet our pages just finished confirm what we broached in the earliest, that the Union more than the Confederacy stood on the tactical defense in the campaigns of 1862 and 1863, and despite superiority of numbers, usually suffered defeat. The reason: a passive defense and a nervous, fearful wait for the enemy, the Confederates, to act. And when they acted, they surprised, flanked, and outmaneuvered the Federals though the Federals stood on defense and enjoyed the favor of time and place. In short: an alert, clever, nimble enemy drew them out of advantageous positions and from behind fortifications and into places and stances where they could be hurt. Let me illustrate:

- Before Richmond, June 1862, the Federals enjoyed the Chickahominy's broad, swampy mass as a shield. Lee took his forces out of Richmond because swamps shielded it as they did his enemy, McClellan. Lee went around McClellan's right at Mechanicsville. When Lee had deployed himself and engaged McClellan, Jackson arrived from the Shenandoah to hit McClellan's army in the right flank in the next day's battle, Cold Harbor. Lee and McClellan had been on the defensive. Lee thus left it and seized the initiative. Meanwhile McClellan persevered, motionless and dogged, on defense. During that time, Stuart had taken his ride around McClellan, circling that army at the rear.

- Jackson even more tellingly enveloped Pope at Second Bull Run, August 1862. Jackson stalled and fixed him by marches, countermarches, feints, and the like. He had Pope where he wanted him at Cedar Point and circled him in a wide sweep. Only an accident saved Pope from disaster. Jackson would have wiped him out.

- Enveloping and encircling at Chancellorsville, the Confederates hurt Hooker, too. In one respect the case differs from the one above. Without advances and retreats to confuse Hooker first, and without a hint, Lee left the secure position at Fredericksburg, confronted Hooker's army in its forward motion, and directed Jackson, barely arrived at this destination, to hit the flank and rear in the Wilderness. The strike spread such terror among Federals that the next day Lee could dare hit them head-on in the vitals and take their position.

- Confederate offensive power did not weaken and tire until Gettysburg. A Union position there, thought weak, proved strong. No flanking preceded the attack. Though in force, it failed. The wizard of the flank attack, tireless Jackson, did not take part. He had been killed elsewhere earlier. The Confederates missed the cavalry, too, the army's eyes and ears. It had been dispersed in a useless raid.[5] The period of offensive tactics, 1862–1863, ended with this battle.

Lee and his Confederates also stood on the defensive now and then. We cannot leave the period without a look. He would secure the front and anchor the flanks: strong all around. Trenches would fortify a front, so arranged at Fredericksburg that a multiple cross fire could be opened. Then, for example at Sharpsburg (Antietam to Northerners), he tried to puncture the enemy's weak spots with offensives that would destroy and damage. Or, deciding against such thrusts, he would keep the enemy under surveillance using the offensive mobility of cavalry at the flanks, as at Fredericksburg. He could block any encircling column or at least be ready by watching its every move. Thus Lee did not receive an enemy passively by waiting to see what he would do. Lee waited on the que vive like a beast of prey, as if at rest but every muscle taut, ready, set to spring.

## 4

In the last and least-known period, Lee conducted a defense and applied defensive tactics. He, like Napoleon in the retreat of 1814, displayed his talent for generalship in full for the first time. Never had he led so well before. Three years of study and experience developed his capacity for command. Fulfilled by combat-tested troops, it shone forth. These operations, these tactics, these battles ended in the collapse of the Confederacy, then? True. But he led an army that once seemed invincible, retaining glory in defeat, and to any soldier teaching lessons nothing less than instructive.

I have said that the Confederacy's offensives ended at Gettysburg. There the Confederacy threw in the last of its strength and the end of its youth. The Union could muster manpower beyond measure, hordes. Thereafter, if the Confederacy punched holes by bloody battle, Union reserves under Grant's direction could with that general-in-chief's energy plug them and invigorate the army with never-ending forces. Lee must be chary of what remained to him. He had seasoned veterans; he had nothing else; they must live to protect and preserve their country as long as possible. Lee said to me at the time: "We cannot get peace by an offensive war. It's clear to me. The more we win, the more hatred we excite in this civil war. So I'm limiting myself as much as I can to defense, to spare my men!" He had to give up strategic defense; he must return to the tactical. Grant wanted nothing less than to be on the defense. He loathed it. Lee compelled it, waged it so well he forced Grant to imitate him.

On the retreat from Gettysburg, Lee chose a strong site on the north side of a Potomac too swollen to take his army across. Even that early he remarked to me as we rode along the fortifications: "In the rest of the war I want only to be attacked in such a prepared position." He confessed with a joy unusual, striking, in the circumstances; such being his faith in that strength. He said he rejoiced that, despite the rising river, the Federals might take the offensive and give him the battle he wanted. Meade has been often reproached for lackadaisical pursuit of Lee after Gettysburg. But not only did Lee stand in a strong place; morale also remained high in Lee's men. Meade would have suffered had he attacked—no doubt about it. The defensive period began then and there.

Tactics newly adopted would be developed and refined to virtuosity, so let us look at them. To an extent, circumstances called them forth. As we have seen in the second period's battles, the outermost skirmish lines began, little by little, to resemble knots and clumps: strong in firepower, and tenacious, but harder to maneuver. Consequently, only nearby objectives could be tried. What happened when such a line found itself in a forest too thick to see through and crisscrossed by creeks and ravines? Leadership of even the smallest units often failed then, because it ended, unable to continue. The men would be on their own. Unable to see right or left, they could make out nothing and determine nil. Worse, on this front that could not be observed or comprehended, the enemy might flank a line or approach a point from the rear, enjoy slight success, and strike terror into leaderless opponents. Fighting in those thickets, how the unled would fret for their safety and fall prey to the nastiest accidents, often fatal!

Lee would let forward lines build light cover, if merely to give something to touch, against those fears. As soon as he took a position he wanted to hold, he ordered the troops in. Fortify the front! Quickly they felled trees parallel to it and, using overcoats and backpacks, covered them with earth. In the open, rails stacked and earth-covered took the place of trees. By rails I mean the crude, stout articles from primitive fences American-style. Such breastworks did not protect from artillery, even worsened its effects, but shielded without flaw against balls and bullets. Behind them no ravage by rifle or musket. To fortify thus grew routine, done in a few hours later in the war. More and more the reserves, some distance back, and here and there the second line, got permission to fortify, too. One became two or three lines of defense. Offensives would debouch from openings left at apt spots for the attack. Behind those openings and to the rear of all, artillery waited, at a distance but to be drawn to battle. Of course, in a forest, artillery ought not be far. Whatever the size of clearing or spread of glade, the gun counts only as far as the gunner can see. In forest fighting, wings and short flanks made the most sense; they could protect one another as well as the forward lines; and they would impart the feeling of security to troops thus protected. Wings at the ends of works would also shield flanks. Strong reserves would add protection to extended front lines. And mobile units, able at once to attack anywhere, would secure the entire position. While the excellent cavalry guarded lines of communication, it would provide information about the enemy. Indeed, let the enemy lift a finger, the cavalry would report it. Indeed, without this intelligence, defense of this intensity would be impossible.

Lee let the Federals run at those fortifications. With the blaze of musketry, usually, and the hail of cannon, he met the assaults: barrage after horrible barrage. He let them soften themselves by impetuous attack and consequent injury and fatigue. Then he turned their woes to his advantage. He threw his men against the numerically superior but physically and emotionally weaker enemy until the enemy did not want to attack anymore. The enemy tried time and again to storm such positions, bloody attempts and vain, before they saw what had been forced upon them. They took up their enemy's way of battle then: defense.

Lee thus dealt with Grant's first thrust in the Wilderness in May of 1864 but did not perfect the tactical defense until the struggles of Spotsylvania Courthouse. Grant—later in May, at Spotsylvania—slammed at him the same way as earlier. They knocked each other around for eleven days straight. Grant's left advanced. Lee saw that it would draw him out of position. He also felt the fatigue of the endless tussles. Suddenly he led with his left and hit Grant's flank and rear. Meanwhile, using the right to seek a new position, he established himself on the Anna River. There the contest continued, a defensive match of shovel and axe. It bored Grant; he loved attack, the bolder the better. At Cold Harbor, May 31 to June 12, he lashed out with the power of four times as many men as Lee's, intending to end the war at a stroke.[6] In tight formation the army advanced with the crush of concentrated assault. The armies butted and slammed for an hour and a half, the war's bloodiest. The concussion ended with them stunned and battered, spent, unable to continue, back where they started.

Lee had shackled his stronger foe in the chains of perfect tactics. Never did Grant try again to shake them off. He accepted—and both generals fought—a defensive war. Each used defensive tactics with a thoroughbred's temper: the bit between the teeth, a leap forward whenever terrain and circumstance permitted, the enemy and victory their objective pursued with a vengeance. I mean especially Grant before Petersburg: grim, tenacious in the battles there. Both leaders undertook little on a large scale; they fought local battles; they damaged one another's lines in brawls and slugfests over bits and pieces; they advanced inchmeal; they withdrew at a snail's pace. What they captured, they secured: the enemy's cover often turned to the victor's use. The loser, driven out but not far, would modify his formation accordingly, perhaps even let his second line take its chances under fire. Thus the last battles resemble a siege, a war of shovel and axe. Hostiles glower at each other eyeball to eyeball day and night, never safe from assault, movements and maneuvers stalled while armies expend strength gasp by expiring gasp in struggle after ongoing struggle for tiny upon miniscule objectives the extreme of defensive warfare. Assault means neither offensive nor charge anymore. The attacker gains his bit, fortifies at once, and creeps from this to another success like up a ladder rung by rung and clinging to each, this step as small and as bloody as the last, to be continued until the foe be lamed and yield. We know who yielded at last.

## 5

The American way of war seems so strange, then, so odd according to Europe's military science now. Let us compare the American to the German for the last two periods. I want finer points thus to be illuminated.

Bear in mind that over here we know nothing like American terrain and consequent troop disposition. Forests cover two-thirds of Virginia; they above all influenced every tactical configuration. Yet, even adding to them the differences between the American then and ours now, similarities in tactics remain to impress any soldier. The Americans abandoned the column as a combat formation, elaborated the line, even multiple lines; and those resembled ours. So did the strong ones of skirmishers out front. We miss nothing but what the French call *soutiens,* the supports and props that stiffen like joists. I think the Americans intended that the first line replace them. Tightening and stiffening by the first, for a thrust, instinctive with the Americans; and the pause by skirmishers: together to strengthen and readdress positions. We do these things as the Americans did. Likewise if the skirmishers, often from the diversest of units, obeyed the man in command by accident of seniority, they obeyed their own inclination by the intuition of robust, resolute nations and their warriors who know victory: the instinct that says, Forward!

Like ours, their formations of course intermingled at the front. As with us at Wörth and Spicheren in the Franco-Prussian War, the Confederates had no reserves in most engagements; to gain the objective they threw every man into battle. Here and over there the absence of reserves meant that when it reached crisis, the commanding general lost influence. Upper levels of command deployed and set in motion an ingenious array, but field and line officers must direct them to success. Panoramists create the designs; miniaturists must fill in the details.

The American army differed from the Prussian in a way no less than profound. In the American, some commanders down to the brigade had been trained as officers and gained savvy with the regulars; and some had distinguished themselves by talent, aptitude, and other personal strengths. The same cannot be said for line officers, especially captains and below, where quality diminished to doubtful, and knowledge degenerated to ignorance of military things. As with us in the days of Frederick the Great, the Americans might treat the brigade as the tactical unit. Into it then—rarely into the regiment—they articulated smaller units, and into that tight organization tried to diffuse the influence of those lower officers.[7] Hence stiffness in the lines and clumsiness in management and direction of troops. The loss of an upper-level commander would cripple advance and retard again in battle.

Contrast Prussian and our brainy line officers, a fact to be heeded. The new Prussian tactics, adopted in 1808, freed the luminaries to use their own minds. This independence of thought not only limbered the stiff lines but also turned loose what had been imprisoned in the toils of form. Liberated battalion and even company commanders could be the heads of tactical units, their own, and make them fight as right-thinking officers saw fit and as well-trained troops best could. The flexible line at the forward edge resembles a chain then, with detachable links under independent guidance. At crisis they can dismember into smaller and even the smallest units without dysfunction. Separating into smaller, discrete units, we Prussians animated and invigorated even the smallest. Our Prussian tactics thus gave our line officers energy, elasticity, and speed—to the entire army's benefit.

Ours therefore exceeded the Americans' in the productive use of lower-level commanders, especially in its transformation of their units into those of maximum effect. Furthermore, diligent peacetime training provided our troops an abundance of formations, something to fit any circumstance; and such proficiency in assuming them that even in battle's fire and chaos, they could assume any without flaw. Lee, the first American to acknowledge this superiority, replied in the thick of Chancellorsville when I spoke with amazement at the bravery of Jackson's corps: "Just give me Prussian formations and Prussian discipline along with it—you'd see things turn out differently here!"

Of course, as we Prussians gained by divided and shared leadership, so we lost in other ways. Despite the severest discipline, any hydra-headed system must lead to fragmentation and a weakening of thrust overall. Nothing but strong reserves can stiffen and support such a ragged colossus.

American tactics of the second period suffered some of the same ills: fragmentation of thrust, intermingling and confusion in the outermost lines, and a consequent breakdown of coordinated leadership. We over here have tried these remedies:

- Care and precision in training the soldier to be independent and self-sufficient, that he act right and on his own, whatever the countermeasures, above all at the moment of decisive clash;
- Training him to changes of direction and shifts in position, including ones to the rear, and to orders from officers he does not know, and to assume on call any formation at once;
- Training the company to independence and self-sufficiency, too, it being the unit that can exist on its own and direct itself while being autonomous enough to carry out the coordinated, single-minded, centralized intentions of the upper-level command; and
- Enforcing one rule on the offensive: overpower the objective with a front of fire upon fire.

Every German officer has also learned what it takes to train in those difficult and contradictory principles: the strictest personal discipline as well as rigor and precision in the company's drills and maneuvers.

America did not have time at peace to train for war, could not learn how to teach those objectives and then reach them. Such objectives demand professional leadership, furthermore, a corps of officers grounded in soldiering and devoted to it. The new tactics for firepower caused injury, damage, and losses. To cope with ignorance and inexperience, the Americans developed these expedients that helped, as we have suggested:

- The soldier and the line officer took predetermined positions, following instructions that made everything clear;
- They got time to think things through, rather than risk confusion by rapid intricacies not yet mastered;
- They erected breastworks as safeguards against panic no less than as shields against enemy fire, to allay the fear that ravages green troops; and

- They must use shovel and axe, exploiting the skill every American has with those tools.

Battle-tested generals instituted them. In time, intelligent, veteran troops carried them out. Would they be of no value in our tactics? We, too, can find ourselves having suddenly to fight on several fronts, as in the Seven Years' War, Frederick the Great's war. In his campaigns we have the finest model. He conducted masterfully those defensive tactics against enemies of superior numbers. We would have to fight accordingly—here block with weak forces an incursion, here drive out the enemy with a bold and resolute offensive and, because weaker, hitting him the harder. Even in 1870 we knew the touch and go in spots now and then, the question of whether our defenses could bring that combination to bear. We ought especially to consider American tactics when parrying thrusts on remote fronts and with older conscripts weak in the aggressive impulse and far from their time of training.

Let us not overlook who took these measures and used these tactics: two generals, Lee and Grant, famous for energy without mercy and without stopping: generals renowned for the urge developed to the rage to *attack*. They saw themselves forced at last to what we have described. The tactics proved reliable; opposing ones failed. Of course either general would have gone wrong with them had he felt superior to the other in a fair fight, and had the other not forced him to waste the time they demand. For, unless circumstances dictate what the circumstances dictated there, the way of war has been and remains the initiative, the assault that risks everything, the aggressiveness, the savagery that orders up the last reserves and throws them in. As long as the commander thinks the strategic chances favor him—and he can keep fighting until the enemy can fight no more—and he can then gain the goal, that long will he attack. When and where necessary to delay an enemy of superior numbers, however, or to save with weaker forces one wing at the expense of the other, the advantage lies in defensive tactics.

They can defeat a stronger foe nonetheless, if used with skill and adapted to circumstances. Of course such adroitness, the application that wins on defense, demands a leader like our Frederick. In the extremities of defense, his eye he never failed; he spotted each enemy's every weakness, however slight. He stayed on guard always, ever applying the resilient verve that snapped his troops out of the lethargy so readily induced by the postures and attitudes of defense.

At war's end we see the Confederacy leave the defense one last time. Mistakenly, unfortunately, the Confederacy's Lee takes the offensive in mountains that favor his small forces but defensive tactics; and that still offer sustenance because not a land trampled and ravaged, but all the more where a defender may defend and defend. Lee, however, a campaigner rather than a defender, will be true to his nature in the final battle. Compelled too long to defend, he will not be boxed in and choked as at Petersburg but face the enemy in honest combat. He shakes off the chains. Once more in desperation, he, our noble Lee, battles Grant toe to toe: ready and rather to go down in a place freely chosen and dearly won: in forests where his battle cry so often echoed and resounded.

# 3

# Infantry

*In America the organization of both armies resembled the European and especially the French.*

### Recruiting—Officers—Organization—Formations—Arms.

The North raised infantry in two ways, the call for volunteers and the voting of bounties to attract recruits. By the end of the war they had become a combination of volunteering and the draft. The South, out of love of country and a small population, expected universal service. It wanted to raise every man; even then the total would be few, compared to the North's masses. It had to raise every man; it waged nothing less than war for hearth and home.

Northern forces proved superior in the mass; the Southern, however good man for man, could not stem that tide. Indeed, man for man, Southern proved superior. Only in the West did troops on one side match those on the other, man for man. Elsewhere the methods of enrollment, the kind of men, and the intensity of commitment produced opposing forces of different elements. The South had neither much immigration nor a laboring class, unlike the North. It followed that city dwellers mattered in the Northern army: factory workers among the rank and file, lawyers among the officers. Lawyers of course understood better what to do with the quill than how to deal with terrain on the march and where to send troops in combat.[1] The Southerner, mostly of the planter class, grew up outdoors and loved hunting and sports. Used to bossing Negroes in numbers, attending to their needs, and making them obey, he brought to the army the officer's qualities. Other factors aided Southern battlefield excellence: the rebel's austere Old English upbringing, for example, and his devotion to the Cause.

## 2

Companies elected officers, Union and Confederate. Had the governments wanted to drop that system, they could not suddenly and at once have found enough with personalities for leadership. They would have had to commission incompetents. On both sides, and especially later in the war, the governments named the line, staff, and general officers. In the early years a man might raise a regiment, brigade, or other unit by advertising and other methods and be named its commander. Later, many staff officers still got appointed out of civilian life and on their own initiative. In the second year both armies established boards, a corrective to elections. Boards would review lower-grade officers who showed a lack of skill, ability, or knowledge, and discharge the inept.

Regular officers, before the war, enjoyed a scrupulous education at West Point. It would count for the Confederates that over two-thirds of them had been born in the South. As most of them sided with their states, so most of the officer corps went to the Confederates. At that time the regulars consisted of nineteen regiments of three battalions each, but few Southbound officers cared more than a little that the rank and file stayed with the North.

On the appointment, tenure, and promotion of officers, I refer to a passage in the *Report of the Committee on the Conduct of the War.* It speaks worlds about the subject on both the Union and the Confederate side.[2]

> *Question.* "Are our troops as well disciplined as those of the enemy, as a general thing?"
> *General Gibbon.* "I do not think they are."
> *Question.* "What reason do you give for that?"
> *General G.* "[After a few preliminary words.] A great deal of it depends on the kind of officers they have. I think we ought to have some kind of rule for promoting officers for efficiency and gallantry in the field."
> *Question.* "Have they any articles of war, or rules of proceeding that will enable them to discipline their men better than ours."
> *General G.* "I do not know that they have. But I believe that all their forces are under the control of the central government. I do not know how it is, but that is my idea. I think the central government makes all the appointments, but I will not be certain of that. But I am satisfied of one thing: that men in their service get their promotions from military recommendations. In our service, the governor of each State has the appointment of officers from his State, though he appoints none higher

than colonels; and very frequently they will insist upon appointing men who are not competent in a military point of view. The general government, however, has the power of dismissing officers; that will enable us to get rid of the worst officers in time, but we have no system which will enable us to get the good officers in position, and sometimes the bad ones are put back upon us."

*Question.* "The rebels, of course, must have selected many of their officers from civil life, as we have done?"

*General G.* "Yes, sir. Do not misunderstand me. I do not want men of military education, but I want men who have shown themselves in the field to be competent. There are two kinds of military ability—one theoretical, and the other practical. We have plenty of volunteer officers who have shown themselves to be really competent soldiers, but they do not get promotion. [He continued with an example.]"

*Question.* "Does not that arise from the necessity of having rules by which it is rendered very difficult to supersede an officer? . . ."

*General G.* "We have broken that rule during this war. Suppose we had promoted in the regular army altogether according to the rule of seniority; you would not have had one officer who has distinguished himself in this war, unless you had promoted all the older officers and then retired them.

"This is the rule which I say will give us a good army, the best in the world: We have the best material upon the face of God's earth of which to make an army; I am convinced of that. In the first place, let us commence with the regiment. After a regiment goes into service, if the colonel is not fit for his place, put him out, either by a court-marital, which is slow, or by a board, which a great many people say is unjust; and then put a competent man in command of the regiment; and then never make a promotion to a captaincy or field officer in that regiment unless upon the recommendation of that colonel. And as long as that colonel tells you that this or that captain of his regiment is a competent military man, never make a promotion in his company except in accordance with his indorsement. If the colonel tells you that the captain is not a competent man, and insists upon having somebody that he knows to be a competent promoted, take the colonel's word and promote accordingly."

*Question.* "Your colonel may have his favorites, and dispense his recommendations accordingly?"

*General G.* "That is true. But how are you going to avoid these partialities? Who is going to decide upon them? Suppose the colonel is an indifferent military man—is not fit for his place; he will have his prejudices as much as anybody else. The only question is, whether the competent military man will not be the surest source from which to make those appointments, who will make them with an eye solely to the good of the service, and will be a more competent man to judge than a man who is not declared by military authority to be the right man for the position. In both cases you will probably have partiality; but in the one case you have partiality governed by military principles; in the other you have partiality governed by politics, relations, anything else, you do not know what."

Then the general related several instances to illustrate what he had said. He praised two Wisconsin regiments; promotion had been turned over to them. He mentioned on the other hand that before the battle a court-martial had sentenced one of his regimental commanders to be discharged, but he had been sent back to him, reinstated, as soon as possible "by the governor of New York." He related further an example of the commander of an Indiana regiment. He could never use him; he reported himself sick, though in an important time of the campaign, yet in spite of his, General Gibbon's, remonstrances, not only returned by official order but also with the brigade commander's commendation. When he, Gibbon, objected, he got a curt answer to accept the man because a higher-up would have nothing else.

### 3

In Europe the French had not yet lost their martial prestige. In America the organization of both armies resembled the European and especially the French, with the regiment the tactical unit.[3] The regular army's regiments comprised three battalions; the volunteer, ten companies of one hundred men each: one thousand to the regiment. The company consisted of two platoons. Two companies formed the unit commanded by the senior of two captains. The seniors of those pairs commanded the odd-numbered and proceeded ahead of the even: 1, 3, 5, 7, 2, 4, 6, 8. Numbers 9 and 10 served as skirmishers as well as 1 and 2 posted in rows at the flanks. A line officer, major or lieutenant colonel, commanded each wing of battalion; a colonel, the regiment. These took post behind the line or front.

### 4

The Union army followed Casey's tactics.[4] Let me mention the regimental formations; regiment being the tactical unit, as I said above. An overview suffices.

(1) The line. Eight companies to the front; skirmisher companies behind those at the flanks.
(2.a) Double column at half-distance; columns formed by companies.
(2.b) Column in divisions.
(3) Battalion by division columns.
(4) The square, empty inside. Though the manual prescribed it, the army did not use it even on the drill field.

Thus I should also say here at the outset that as the table of organization, the scheme of formations the order of battle, and the manual of procedure followed the French, so the American—like the French—seldom used in battle the rigmarole of peacetime and the drill field. Units in combat became independent; tactics, autonomous. Indeed, consider the formations listed above. A split into two wings, each under an independent commander, ought to occur; in battle it practically disappears. Unfortunately, textbook exercises amount to drills and shows never used for real, unless something of them remains in tactical elabortion in combat.

The brigade consisted of two to four regiments of infantry, a battery of artillery, and one thousand cavalrymen in four squadrons. Soon after the war began, cavalry belonged to the brigade only on paper, having formed independent units. Three to four brigades made up a combat division; and two, three, or four divisions, ten to twenty thousand men each, a corps. The corps took the order of battle sketched.

The Confederates, following Hardee's tactics, used about identical formations to the Union's.[5] If they differed from Casey's they minded even more the example of the French.

Troops at drill would exchange parade for battle formation when passing in review. I saw it at recruit depots in Richmond, whence replacements to the armies in the field; and in camps, where in winter they drilled as if they meant it, to the beat of a big drum. Indeed, they stressed this transformation as an essential exercise. Consequently the Confederates, iron in march-discipline anyway, formed battle lines with incredible speed, and complete even to skirmishers out front.

March-discipline meant rigor up to the final minutes and the commands for battle order then. March-discipline meant distances observed with care and skill in keeping to formation. Companies obeying march-discipline must maintain integrity with cohesion into the tightest possible composition. Thus the army kept together within units and one unit to the next. The army held to tempo and maintained the cadence, too, the severe, unrelenting pace to cover the most distance regardless of terrain. The Confederates in their armies of astonishing marches averaged ninety to ninety-six steps a minute. Jackson's troopers, the so-called (with a chuckle) foot cavalry, also obeyed this tempo on foot.[6]

Drills included marching in open columns and single- and double-company formations at various intervals and distances, and with wheeling or turning (evolutions), and assembling or massing, and passing in review. Forming to fight meant nothing but deploying in line, rows of skirmishers out front.

5

The infantry went through stages, arming. At first they carried old, even obsolete weapons of all sorts, out of stockpiles and bought in Europe. Plenty of Miniés and Podevils, for example. Confederates even carried single- and double-barreled muskets. Later the North adopted Springfields, a twin line to the British Enfield. Only sharpshooters got rifles with telescopic sights for distances and precise targets. The breechloader found little favor in equipping the green armies poor of discipline and wanting in marksmanship. The many battles in forests increased dislike for the breechloader. Accuracy of the single shot—not the allegedly smothering barrage—won in forests. Lee and I once talked at length and in detail about the difference. He had armed a brigade with breechloaders, a trial. "In an hour and a half the men had already exhausted their ammunition. Back they came from the front. We cannot manufacture so much, nor transport it, unless we get results. I strive to cut its use to the least. We need a weapon that demands time to load, so the man knows he must value the shot—not fire before he directs it to a consequence."

Except in the West, in sum, Southern infantry, though always fewer and in the later years diminishing, fought better than the more and more, and ever more numerous Northern. Only in the West, forces being equal there, man for man, did victory after clear victory go to the side with more men or better leaders.

# 4

# CAVALRY

*Cavalry meant and must mean—first and foremost—warfare by shock.*

Introduction—Of the Confederacy—Recruitment and Organization—Tactics—Two Examples from Von Borcke's Memoirs—Stuart and His Lieutenants—The So-Called Raids, with Examples—The Union Cavalry—Comparison of Confederate and Union with European—What Raids in the American Style Can Teach Europeans about the European Style.

**C**avalry fell into disfavor in Europe after the wars of 1854 and 1859.[1] The head of our armed forces raised doubts—people questioned what he had done—when he enlarged the cavalry and increased its preparedness. Precision firearms had ended cavalry's dazzling career, theorists said. Armies would tolerate mounted regiments as anachronisms, of course, something to remind them of the past. But "ten infantrymen can be maintained for every saber put on the field. Ten of foot will accomplish more besides, much more, than ten of horse." This, the verdict of Theory, we could read in military journals and nearly every newspaper. Military history, on the other hand, confounds the most brilliant theories. Against the theorists, the martial schoolmaster, History, proves *his* lessons with facts and beyond doubt.

The American war broke out. Its beginnings caused scant expectation for anything significant to military science. Still, here and there, something would happen to catch Europeans' attention, win their respect, and increase their knowledge.[2] In cavalry a grand theater and unprepared antagonists favored innovation. Given those conditions, Americans might develop a cavalry of and on its own. Yet the American conditions did as much to hedge in the cavalry as to set it free: thickets of forests crisscrossed by fences and crosshatched by creeks and ravines in a landscape defined by rivers. Nevertheless, European armies and in particular the cavalries noticed, gradually but increasingly, things peculiar to this war but applicable to warfare. As American technique improved, the American cavalry gained attention. Soon its exploits and accomplishments made the cavalryman's heart beat faster. With sympathy, with joy, backers of cavalry watched a spirit of cavalry grow, indigenous and pure, untainted by its own fears or others' reproaches. The same eyes saw develop the genius for cavalry leadership. Cavalry's achievements became manifest, cavalry's prospects the brightest.

Native purity included tactics that matured in no way like models in military science then. Theory posited a dismounted cavalry: an infantry that rode to battle and fought on foot. British authors, among them the brilliant Colonel C. C. Chesney, have written books about this war; and they evidence study and understanding.[3] Endemically, the British favor the dismounted. Shock action does not fit the British countryside, ipso facto the British mind. What nation can develop true cavalry, men who fight mounted, in a landscape that does not invite them? America's spontaneous cavalry took up arms with a will, took to the field with vigor, and took the enemy by storm. Like the Prussians of old, the Confederates and then the Federals put success in the saddle with shock and with steel. None of dismounted cavalry's prudence for those men on horseback!

From the American war the fact emerged: cavalry, including Europe's, had a future. The cavalry therefore deserves our attention. Though an engineer by commission and an amateur at cavalry, I spent twelve weeks attached to cavalry over there. I shall for you, my colleagues, venture to the extent of my knowledge to explain it and describe what it did.

## 2

The infantry of those improvised armies must develop for a time before it improved enough to do its part in them. Cavalry had to develop too, of course, but out of different circumstances. Southerners before the war loved to ride as sport. In addition, farms being apart and roads bad, Southerners traveled distances on horseback. After the wars in the Crimea and Italy, theorists had disdained cavalry here. But over there the Confederacy, a nation of horsemen, would try their luck with it, theory be damned. Confederate cavalry throve, theorists' dislike notwithstanding. Its victories flourished, not so much in pitched battles, though, as in skirmishes and engagements, strikes, ambushes, and coups, small affairs and daring capers, won by horses' speed and riders' courage. In the early days of 1862, when tactics had only begun to take shape and remained rudimentary, Ashby led and escorted Jackson in the Shenandoah, and Stuart rode around McClellan. Ashby and Stuart caught the attention of the military even in Europe. Thus, Southerners first (all the more to their credit) brought honors to cavalry in modern warfare.

The North had less aptitude, inferior inclination, and horses that weighed too much. Still, as the South led, so the North must follow. Little by little it developed a cavalry of firearms and thereby a soldiery of dual roles.[4] Southern cavalry, superior in tactics and better fighters but of shrinking numbers, could prevail no longer. Northern won at last because it had more men.

Northern won in the end, but let us look at Southern first. It possessed the spirit, and the spirit possessed it. Besides, I know the Southern—I served on Stuart's staff. More than once I watched Southern inspected and saw them over and over in the heat of battle. Speaking from experience, I must at the outset cite something emblematic. The American, though handy with firearms, has nothing but thumbs for the sword. The Southerner did better in the early clashes when the green Northerner could barely stay on his horse—better because the Southerner, as Southerners told me again and again, sheathed sabers, seized the Northerner by the neck, and dragged him to the ground. Even in my time in the Confederacy, 1863, Confederate cavalry, for all the emphasis on shock and steel, excelled at shock but fumbled with steel.

### 3

Confederate cavalry recruited its troopers from citizens obligated to serve. The recruit provided his horse, broken to the saddle and ready to ride. A peculiar system replaced those lost in service. A commission assessed the value as each entered, and levied the sum against the owner. A lost one's owner got six weeks' furlough and the money for a replacement. If, furlough up, he returned without a horse, he became an infantryman. Cavalry therefore consisted of well-to-do farmers, an elite who loved to ride.

Devising a scheme of organization to fit itself, it subdivided according to nature and function.

- *Regulars.* Under Stuart, Earl Van Dorn, and John H. Morgan. An element of the army and at the disposal of the general commanding the theater. Organized like European and into divisions, brigades, regiments, squadrons; sizes variable depending on exigencies: squadrons sixty to a hundred horse, according to losses; regiments, four to six squadrons. Chief weapon, cold steel, though per regiment one or two squadrons with good carbines besides the saber. Saddle, the so-called McClellan mode, a compromise between flat and sawbuck, heavy but popular; stirrups, wood, leather-covered against mud and dirt, lightweight but firm of support. None but the curb bit, without bridoon, usually sharp. Buckled fast to the saddle, between seat and upper girth, the saber could not flap and rattle; it also left the trooper free of a weapon that would cramp him fighting on foot. Carrying the carbine fastened to him, however, gave him a firearm to hand when dismounted.

- *Partisans.* So-called free rangers or partisan rangers; volunteers; gathered around a well-known leader. Numbers fluid. Independent of regular army, fought on own regardless of operations. Unpaid; lived on booty: horses, provisions, weapons sold to the government, hence their wages. Best-known leader, General John S. Mosby.

- *Couriers.* Also called orderlies. Mostly young, adroit, resourceful men on excellent horses with minimum baggage. Apportioned to duty in various headquarters; by regulation, sixty to the commanding general, twelve to the corps, six the division, three the brigade. Carried their provisions only when with their headquarters; provisioned first, before the men of the units, when on assignment to other units.[5] Often worked in relays. Carried orders and other messages always in writing; every commander kept for them a pouch of slips fastened to the saddle.

- *Scouts.*[6] Often from Indian country; attached to the regular army now; assigned to reconnaissance, patrols, and spy missions. Picked men; must be bold, resourceful, reliable, and know geography and be savvy about terrain. Usually rode by night past enemy outposts; hid by day with sympathizers or in forests.

On the Pennsylvania campaign, as soon as Lee crossed the Potomac, he had the Union newspapers a day before the Confederacy's —daily without fail—so well did scouts work. Thus did scouts earn respect like few others of the army and know personally almost every officer of the units served. To a man the army mourned the loss of a scout.

Mosby, of Mosby's Partisan Rangers, operated mainly in enemy territory, and mostly in the Union's west and northwest, marauding by tactics Mosby developed himself. He led from 250 to 500 men; arms included a few ultralight cannon; but speed, surprise, and a daring élan numbered in his forces. Fast horses hurled him as if out of nowhere and upon the foe. The attack would be on foot in a loose, single line of troops, its members skilled and practiced to keep it intact even when moving snakelike. Usually he enveloped the enemy in the maneuver therefore called the lasso tactic. Alert patrols meanwhile secured his flank and rear. In his area of operations he had in his favor almost always an enemy of green troops. Should he see a threat to his safety—superior numbers, say—he would disappear like the wind, to reappear elsewhere to strike with speed, surprise, and his daring élan. We regret the lack of accurate reports; they would make thrilling reading. I know from experience that a dispatch from him would raise joy at headquarters.

### 4

The South developed tactics by experience. The Southerner, familiar, at ease, and authoritative with horses, decided to put the mastery to use. His horseflesh stood him in good stead. He had bred the hunter, superior to the Northern trotter in agility, endurance, and speed. So he created cavalry and resisted fighting on foot.[7] Yes, forested thickets, cut through by fences and ravines, together with skill in firearms, tempted the mounted to dismount. But he despised fighting on foot. Sometimes unavoidable, in his eyes it would be ever inferior. On horseback he would be flexibly mobile. From the saddle he could loose the savagery of collision and impact. To him, cavalry meant and must mean—first and foremost—warfare by *shock*.

A superior horse artillery, added to riders who could fight on foot if need be, completed the Southern cavalry and made it an arm in its own right and to be reckoned with. It must be the eyes and ears and at the same time the veil that hid the army's movements and kept the enemy in the dark. Lee demanded: he would accept nothing less: the cavalry must know the enemy's whereabouts at any moment; strength, too; and if possible, intent. The assignment demanded everything the cavalry could muster. Lee took the enemy by surprise and at the right place, and with force that hurt, and won the experts' admiration, thanks to what the cavalry did for him. He accomplished the feat though numerically inferior, because the cavalry never failed him.[8]

I speak with Stuart's exemplary division uppermost. Stuart before action did everything possible to prepare for success. He learned all he could through scouts and spies. He, with staff, took his own reconnaissances, often beyond outposts. The staff included engineer officers; he had them ever supplementing and improving the treacherously unreliable maps. He informed himself about roads, fords, and the like: nature and condition of each. In short he took every pain and precaution, missed nothing in the strain to assure the outcome. Therefore he spent nearly twice the time his men spent in the saddle. Success crowned practically all his operations, thanks not to luck but to unceasing, vigorous, self-sacrificing work, *his* work.

Plan laid, goals determined, he followed it to get to them and with resolve. He succeeded by motion: unending, unpredictable in its independence, and incalculable in purpose; and by surprise: the attack that took unawares because sudden and unexpected, and all the more successful because dogged. He tried to make himself feared for the attacks. So without fail he struck where least expected. He tried to cut off troops that had lost their way or been otherwise detached. He loved to demolish depots and visit panic upon enemy communications at the rear. He never went around an enemy without doing what left him famous, the impetuous, intrepid attack: attacking what he loved to attack, the flanks, and branding them with sharp, resolute steel, his trademark.

To add detail, let me trace with emphasis on cavalry a battle in tactical elaboration. The battle began when the attacker tried to get as near as possible without being discovered. If surprise failed with any slip of the approach, artillery would start. Thus they attempted to force the enemy to deploy and, because Southerners disliked moving under any fire but of their choice, to give the regiments time to move up to attack. If the cavalry faced fortified infantry, or the enemy's whereabouts remained in doubt, squads of sharpshooters would dismount and gesture with motions calculated to provoke the enemy to reveal himself. Artillery of course helped spot him. Many consider such cavalry contacts to be infantry engagements. I have indicated that Southern cavalry did not fight on foot. How much did they dislike playing infantry? In skirmishes of quick development, squadrons pulling out in a hurry would go so fast sometimes as to abandon the sharpshooters.[9] Cavalry on occasion did dismount, but only, like the artillery, to give the army time to draw up for battle.

The battle continued with cavalry usually seeking to be at the flanks: their own army's for defense or the enemy's for attack. Against attacks on oneself, one kept ready a small reserve or, as in a few instances, rallied from among the force enough and in time against the lateral threat of a counterattack. When not disposed for flank action, riders assumed two lines within the brigade, then joined the infantry in the attack. Attack begat the moment of truth. Cavalry formed into the double row, the second a horse's length behind the first, a formation neither tight nor constricting. So arrayed, they struck. At the point of charge they let out a bloodcurdling yell like that of infantry with the bayonet. With force, with resolution, with grit they struck, they hit, they slammed. War to them meant warfare by *shock*.

Yet—to shift from description to critique—we cannot blink the fact: Confederate cavalry could not decide battles. In this obstacle-replete terrain, nothing larger could maneuver as a tactical unit than one too small for decision-making: the brigade. Even when a terrain would have let a larger be brought to bear, smaller seldom united to be larger. True, Stuart increased his corps to twelve thousand with brigades cooperating toward one goal after a single plan; but each remained unto itself in tactics.[10] Besides having to be big enough and coherent, cavalry, to be decisive, must sustain its attack. So much the worse for the Confederate. It lacked persistence because undersized and disjointed.

The use of echelons would have stiffened and sustained them in their independent, self-contained state, perhaps made them more influential. Now and then, regiments did attack in echelons. At Brandy Station, for example, with success. Stuart and Von Borcke accordingly wanted cavalry left status quo and to individual calculation and prevailing circumstance, not fixed by regulation to contradict such success.

I have just mentioned Von Borcke. I return to him in the next section, to illustrate and finish what I have begun in this. Details from another viewpoint will interest you.

## 5

I borrow from Von Borcke's peerless memoirs. I insert verbatim the discussions of two engagements: the assault on Brandy Station, August 20, 1862, an attack that followed the rules; and the sketch of the raid two days later on Catlett Station.[11] They illustrate what I said in the prior section.

"*20th August.* — At daybreak, with two brigades, we crossed the Rapidan. The passage was attended with difficulty, especially with the artillery, on account of the depth of the water. Lee's brigade was sent to the right, in the direction of Kelly's Ford; General Stuart and staff marched with Robertson's brigade in the direction of Stevensburg, about one mile from Brandy Station, and both commands were to unite near the latter place. Our advance-guard came first in contact with the enemy, who, broken by the attack, fled in great confusion, and were pursued through the little village and more than a mile beyond it. The joy of the inhabitants, who for a long time had seen none but Federal soldiers, and who had been very badly treated by them, cannot be described. Men, women, and children came running out of all the houses towards us with loud exclamations of delight, many thanking God on their knees for their deliverance from the hated yoke of the enemy. A venerable old lady asked permission to kiss our battle-flag, which had been borne throughout so many victorious fights, and blessed it through her streaming tears, and many a bronzed and bearded face in our ranks was moistened as we witnessed the touching scene. The enthusiasm was so great that old men and boys, all that were able to carry a gun, in spite of our earnest remonstrances, followed our column to join the fight with the detested Yankees.

"The enemy, strongly reinforced, had now taken position about two miles from Stevensburg, on the outskirts of an extensive wood. Several small detachments had been pushed nearer towards us, and were patrolling on our flanks. One of these, in strength about half a squadron, mounted on grey horses, operated with great dash, but, advancing imprudently, was cut off by a body of our men, who fell upon them like a thunderbolt, killing and taking prisoners all but six, who saved themselves by the fleetness of their horses. The Federals dismounted many of their cavalrymen; and their line of sharpshooters, about a mile in length, poured upon us from the dense undergrowth a heavy fire, wounding several of our men and horses. This checked for a time our onward movement; but a large number of our troops, having been also dismounted, engaged the Federal tirailleurs with great gallantry, and we then charged with the main body upon the enemy's centre, and quickly drove them from their position.

"In the *mêlée* I captured a very good horse, which was unfortunately wounded very soon afterwards, but I took from it an excellent saddle and bridle that had belonged to an officer.

"The enemy's retreat was now so rapid that it was difficult to keep up with them, so that General Stuart, in order not to exhaust all our horses, took only one regiment, the 7th, in the pursuit with him, giving orders to the rest to follow at an easy trot. We were not long in reaching the heights near Brandy Station, from which we saw the Federal cavalry in line of battle in the large open plain before it. They were about 3500 strong, and, being drawn up in beautiful order, presented, with their arms glittering in the sun and their waving battle-flags, a splendid military spectacle. Our brave fellows of the 7th were immediately placed to confront them, and the sharpshooters of both parties were soon engaged in a brisk skirmish. With great impatience and anxiety we now awaited the arrival of our reserves, and courier after courier was sent to hurry them to the spot. As even our colour-sergeant had to perform orderly duty, I took the battle-flag from his hands. This act attracted the attention of the enemy's sharpshooters at a distance of 800 yards, and they kept up, from that remote point, for some time, a surprisingly well-directed fire at me, one of their bullets cutting a new rent in the glorious old ensign.

"The enemy now commenced his serious attack, and as our position, by reason of his vastly superior numbers, was quite a precarious one, General Stuart, taking the standard himself, ordered me to gallop in haste to our reserves, assume myself in command, and bring them up as fast as the horses could run. After a short, sharp ride, I reached the regiments, and with a loud voice commanded them, in the name of their General, to move forward at a gallop. As I was well known to every man in the division, the order was at once obeyed, and in a few minutes I arrived with the column at the spot where General

Stuart awaited us with the greatest solicitude, just in time to form hurriedly our lines and dash onwards with the wild Virginia yell to the rescue of the 7th. Occupying the place of honour in front of the regiments, I shared to the full extent the excitement of the onset. The enemy, as usual, received us with a rattling volley, which emptied several saddles; but a few seconds more and we were in the midst of them, and their beautiful lines, which we had so much admired, had broken into flight. . . .

"During the confusion of the *mêlée*, I discovered suddenly that a fresh squadron of the enemy was attacking us on the right flank, a manoevre which, in the disorder inevitable after a charge, might have turned out disastrously for us; and collecting about 80 of our men around me, I threw myself with this comparatively small force upon them. They at once slackened their pace, and when we had got within forty yards of them, halted and received us with a volley which had very little effect. Upon this they fled precipitately, and were chased by us into the woods, where many of them were cut down and made prisoners.

"The main body of the enemy meanwhile had rallied several times, but again and again they had to yield before our impetuous advance, until the last of them were driven through the waves of the Rappahannock, where their infantry and artillery, strongly posted on the farther bank, offered them protection, leaving behind many of their dead and wounded and several hundred prisoners. . . .

"*22nd August.*—The darkness of the night had not yet given way to dawn, when we again set out for active operations, with portions of Fitzhugh Lee's and Robertson's brigades and our horse-artillery, numbering about 2000 men. A strong demonstration was to be made in the direction of Wellford's Ford on the Rappahannock, to divert the attention of the Federals, and facilitate the daring raid we were afterwards to undertake. Accordingly, we marched about five miles northward, crossed the Hazel river, a tributary of the Rappahannock, and arrived about eight o'clock at Wellford's Ford, where the opposite banks of the latter stream were occupied by the Yankees in great numbers. The enemy's artillery was soon engaged in a brisk duel with our two batteries of horse-artillery, which suffered severely, losing many men and horses, in consequence of the superior positions and greater weight of metal of their antagonists. About ten o'clock we were relieved by Jackson's batteries, and, withdrawing from the field without the knowledge of the enemy, proceeding in a rapid trot eight miles higher up the river to Waterloo Bridge, where we crossed it, and continued our march to Warrenton. Late in the evening we entered this little town and were received with the liveliest demonstrations of joy by the inhabitants.

"We were now again exactly in the rear of the Federal army, the right wing of which we had marched around, and our bold design was nothing less than to capture the Commander-in-Chief and his headquarters, which, as our scouts reported, had been established at Catlett's Station, a point on the Orange and Alexandria Railroad. After an hour's rest to feed our horses, we left Warrenton behind us, continuing our march with great caution. Night was now rapidly approaching, and the angry clouds, which had been gathering in the sky throughout the afternoon, soon burst upon us in a tremendous thunder-storm and the heaviest rain I ever witnessed. The narrow roads became in a short time running streams of water, and the little creeks on our route foamed and raged like mountain torrents. But this was the very condition of the elements we could most have desired. The enemy's pickets, in the fury of the storm, indifferent to everything but their own personal comfort, were picked up, one after another, by our advanced-guard to the last man, and we had thus arrived within the immediate neighbourhood of the main body of the enemy without the least information on their part of our approach. . . .

"The rain was still pouring in torrents at eleven P.M., when we came directly upon the Federal encampment, which extended about a mile in length on either side of the railroad. We halted at the distance of about two hundred yards to form our long lines and make our dispositions, which we accomplished without attracting the notice of our adversaries in the heavy rain and amid the incessantly-rolling thunder. The sound of a single trumpet was the signal for nearly 2000 horsemen to dash, as they did with loud shouts, upon the utterly paralysed Yankees, who were cut down and made prisoners before they had recovered from their first astonishment. I myself had instructions to proceed with a select body of men to General Pope's tent, which was pointed out to us by a negro whom we had captured during the day, and who had been 'impressed' by one of Pope's staff-officers as a servant. Unfortunately for us the Commander-in-Chief had, for once, this day his 'headquarters in the saddle'—an intention which he had so boastfully announced at the commencement of his campaign—and had started a few hours before our arrival on a reconnaissance, so that we found only his private baggage, official papers, horses, &c. &c. I obtained as booty a magnificent field-glass, which was afterwards of great service to me. The scene had become in the mean time a most exciting one, and the confusion which is always the consequence of a night attack, had reached its highest point. The Federal troops on the other side of the railroad, which was not so easily accessible, had recovered from their panic, and, reinforced by some companies of the so-called Bucktail Rifles, commenced a vigorous fire upon our men, who were scattered all over the field burning and plundering to their heart's content. In the background our reserves were actively employed in firing the immense depots and wagon-trains and the railroad bridge, and the flames, rising from a hundred different points at once, reddened the dark cloudy night. It was difficult to recognise friend or foe. Shots fell in every direction—bullets whizzed through the air on all sides—no one knew where to strike a blow or where to level his revolver—no one could be certain whether the man riding at his elbow was Federal or Confederate.

"Having received orders from General Stuart to cut out the telegraph wire, I proceeded with twenty men to the execution of this purpose; but just as we had reached a pole, I saw suddenly, by the vivid illumination of a flash of lightning, a whole company of the enemy drawn up in line not fifteen steps from us, and I had just time to call out to my men to lie down, when a rattling volley sent a shower of bullets over our heads. I galloped back to the General asking for a squadron to assist me in carrying out his orders. The squadron was immediately granted. Attacking the Federal infantry myself in front, while Colonel Rosser took them in the flank, we succeeded in driving them farther back. But they still maintained a rapid fusillade, and to climb the pole and cut the wire was a very dangerous undertaking. A young fellow of not more than seventeen volunteered to perform the daring feat, and, using my shoulders as a starting-point, he ran up the pole with the agility of a squirrel; the wire, severed by a stroke of his sabre, was soon dangling to the ground, and the brave boy escaped unhurt, several bullets, however, having struck the pole during his occupation of it.

"About three o'clock in the morning the work of destruction at Catlett's Station was complete, and the order was given to re-form and start upon our return. The alarm had been spread over a great part of the Federal wing, and troops were marching against us from several directions. Our success, in spite of the great confusion of the midnight attack, had been very decided. We had killed and wounded a great number of the enemy; captured 400 prisoners, among whom were several officers, and more than 500 horses; destroyed several hundred tents, large supply-depots, and long wagon-trains; secured, in the possession of the Quartermaster of General Pope, 500,000 dollars in greenbacks, and 20,000 in gold; and, most important of all, had deprived the Federal Commander of all his baggage and private and official papers, exposing to us the effective strength of his army, the disposition of his different *corps d'armee,* and the plans of his whole campaign. Our loss was comparatively small; and after a rapid march, impeded only by the deluge of water still pouring down upon us, and compelling us to swim several creeks which were ordinarily but a few inches in depth, we reached Warrenton, with all our prisoners and booty, at eight o'clock the following morning."

6

In any cavalry, plan and execution so interweave, and the elements of battle change so fast, that in no other branch does the leader's personality infuse and unify everything. In this cavalry, higher officers lived cheek by jowl and grew used to one another—as they should for this or any cavalry to succeed. Take Fitzhugh Lee, Rooney Lee, and Wade Hampton, born cavalrymen. Each knew what he must do to aid the success of all. Each enjoyed intimacy with his staff. Each knew their ways one and all, and could use the man as he should be used. The Lees and Hampton in turn had been Stuart's lieutenants. Stuart, in the company of those officers, so often exchanged views, especially after action, that he had but to outline a plan and they understood.

Stuart in the execution not only impressed his men with his courage, cold-bloodedness, tenacity, and resourcefulness beyond end and without limit. For those virtues came with the man Nature had fitted to lead cavalry, every inch of him. But his looks and presence also won loyalty and fired up his men: the flashing eye, the quick and stirring word, and the irrepressible humor not to be checked whatever the circumstances. Brigade leaders put blind faith in such a commander, the tireless one, the master. To the Lees and Hampton he had but to sketch a plan and they obeyed.

Can I refer to Stuart's officers without mentioning the only German officer in the Army of Northern Virginia? I mean the leader with audacity and a keen military eye, a favorite throughout the mounted forces, Stuart's mainstay and chief-of-staff, Heros Von Borcke. The success of Confederate cavalry owes much to this meritorious lieutenant colonel. He wrote an excellent book.[12] No cavalryman can put it down without satisfaction and enrichment. I have mentioned him before in this book. I cite this—I have used it often.

## 7

I cannot end my discussion of Confederate cavalry yet. How could I without the so-called raids? Federal cavalry later, as well as Confederate throughout the war, raided often and with success behind the lines during pauses in the action. *Raid* means a thrust by 1,500 to 3,000 troopers and a few cannon to destroy depots and perplex and confuse commissariats and disrupt supplies. Recall the one I mentioned above, Stuart's first ride around McClellan. It gave the nudge to McClellan's change of base from the York to the James.

True, with repetition, raids lost significance as combatants got used to them. They then served more as tests of what cavalry could do. But at times they still proved crucial. After Gettysburg, for example, Stuart's raids established that the withdrawal ought not be by the same route as the advance. As in the proverb they lock the barn after the thief has the horse, so (he learned) countermeasures would be taken everywhere after the intrepid column has been there.

Let me turn again to Von Borcke's memoirs. I insert a short episode.[13] It will show you, my fellow officers, how Stuart prepared a raid as he prepared all enterprises: with effort. Von Borcke in 1862 has just joined his staff.

"Towards dusk on the 8th we set out on one of these expeditions, escorted by a half-a-dozen of our couriers, and I soon perceived that our ride was to be extended to a greater distance than usual. It was late in the evening when we reached the last of our outposts, and I was not a little surprised when the General dismissed his escort and desired me alone to accompany him farther. Silently we rode through the lonely wood, whilst the darkness drew deeper and deeper around us, and the stillness of the forest was only broken by the strange tones of the tree-frog, and the melancholy cry of the whip-poor-will.

"We soon found ourselves within the enemy's lines; at any moment we might stumble upon one of their patrols, and General Stuart smiled significantly when he saw me examining the loads of my revolver, and observed that he would not employ firearms except in the last resort, and that in case of an encounter we must make use of our sabres. This ride was strangely exciting to me; now that I have become so accustomed to such expeditions, I could go through it with the most perfect composure, but then I was feverishly agitated, and every rustling bough, every bird flying past, increased the strain.

"After a ride of about five miles we reached a small house, and on Stuart's knocking at the door in a peculiar manner it was opened to us. The house was inhabited by an Irishman and his family, and here General Stuart had appointed a rendezvous with one of his spies in order to obtain an authentic report of the enemy's position. This man had not arrived, so we fastened our horses to the fence and went into the house. Hour after hour went by, and the man came not, and it was past midnight when General Stuart became convinced that some unlooked-for hindrance must be detaining him. Not all our powers of persuasion and promises of money, not even my offer of accompanying him, could prevail on the old Irishman or his son, a lad of seventeen, to walk over to the spy's abode, which was about two miles distant, and near one of the enemy's camps, and so the General and I were obliged ourselves to undertake this dangerous expedition. With the first glimmer of daylight we mounted our horses and cautiously set off on our way. The peculiar repugnance of the Yankees to patrolling at night, and the heavy rain which favoured our enterprise, allowed us to arrive without misadventure at the man's dwelling just as the reveillé was sounding in the camp only 400 paces distant. The spy was very ill in bed, and General Stuart had to dismount and go to his bedside. I felt infinitely relieved when the General, extremely well satisfied with the information he had obtained, swung himself into the saddle, and we galloped back to our lines, where we were greeted with delight by our men, who had begun to entertain considerable anxiety on our account.

"Such rides and expeditions were habitual with this bold General, and we often escaped as by a miracle from the dangers which surrounded us. It was only by this exposure of himself that he could insure the extraordinary success which invariably crowned his expeditions and military operations.

"The object of this excursion soon appeared. Our cavalry force received orders to provide themselves with rations for three days, and on the 12th we commenced that ride round the army of General M'Clellan which attracted so much attention even in Europe."

## 8

We tell a different story, turning north. Northerners did not take like Southerners to the saddle to travel and for sport. History and circumstance had left the North as poor in horses as in horsemanship. Not a product of the ways of the people, then, Northern cavalry took rise out of the need to counter the enemy's success on horseback. By diligence, by industry, little by little against unfavorable origins, a Union cavalry worth notice came about. During the first three years it could not hold a candle to the Confederate. In the end, with greater numbers and illuminated by Sheridan's brilliance, Union cavalry made a difference.

Still we mention Union cavalry but in passing. Compare the true, smart, plucky tactics of Stuart, Hampton, and Fitzhugh Lee: how they exceeded everything equine in the North! Union cavalry preferred to fight on foot, the horse a means of transport, only that and nothing more. Discussing infantry we have said enough about fighting on foot. The Russians have tried dual-purpose soldiery on a grand scale, the famous example. A "cavalry of foot" succeeded somewhat in America's thickets of forests, true. But we Europeans fight in the open, the eye sees all and everything on our fields. We Europeans cannot expect advantage from a mounted infantry or a dismounted cavalry.

## 9

If in what follows I dare consider American with Prussian and German cavalry, I do it to throw light on the American. Besides, as an amateur of cavalry, I cannot offer a professional critique.

I saw Confederate cavalry, born riders, leave the saddle only when wounded. Still ours excels in thorough and precise training in fundamentals, in speed become habit executing movements, in sureness of swift entry into different formations, and even in the turnabout or reversal. Behind ours in those ways, could the Confederate measure up in any but leadership? On both sides of the water, with guts and kinetic power, cavalry hurl themselves against the target. The Confederates hit with more shock. Familiar with arms and devoted to hunting, Confederates outclassed us on foot. The Confederate understood how to seize an important site where his fire would count. This art spared many a rider's life otherwise sacrificed in vain. Confederate cavalry disliked fighting on foot as much as ours disliked it but never hesitated when circumstances seemed to demand dismounting to hold terrain or dislodge a foe.

As to scouting, make no mistake: the likes of it in the American war has never seen its equal anywhere else. I doubt our system lends itself even to the attempt: to finding and training men for nothing but reconnaissance and the gathering of information. Commanders over there got reliable data and the succinct report from scouts, not superficial, vague utterances from any number of patrols, a jumble that often does more harm than good. Picked men schooled for the purpose, serving as outriders and rangers, did not stop at communicating the enemy's size and position. Those scouts differentiated among his corps, and learned and reported how experience had fitted each for battle.

If we cannot match the scouts, we could emulate the couriers. They had the advantage over ours threefold: blooded horses, training as couriers and nothing but, and weapons lightened every way to reduce load. Together they meant the courier could travel distances in the shortest time. We too, however, could find intelligent young men of light weight, familiar with horses, and fit for this service.

We of the European armies lack what might cost too much: understanding between higher and lower commanders. During fighting, a cavalry battle's factors and their interrelations change like lightning. Time will not permit analysis and deliberation, not even the communication of orders. Lower commanders must sense the intent of the upper as if by instinct. This quality verges on an absolute to be acquired over time and on maneuvers when lower observe higher and learn how they behave. No other branch demands it. The grind of battle usually allows other branches to alter the day's orders and communicate changes as part of the regime. Our cavalry's maneuvers might bring together the best and brightest, but a few days' practice will not bring higher and lower eye to eye. Higher ought get the chance to keep in touch with their regiments and brigades.

How, as we look back, does Confederate cavalry measure up in that and related respects? Stuart, its representative leader, has been censured from several quarters. They say he lacked in tactics as to be indecisive in battles that mattered. But the terrain did not lend itself to battles that could be so decided. Furthermore, nobody insisted the cavalry decide. Moreover, undersized cavalry cannot decide. The South could never provide horses for the sixty to eighty squadrons in a theater if cavalry shall decide. Yet anything less might be sacrificed at a bad turn, to become a wall of meat and bones to shelter a clobbered infantry until it recover. Divisions like Stuart's in favorable circumstances can at most be locally decisive but generally irrelevant.

What if a cavalry be rushed together large enough to trample everything in a mass attack? The effort would fall short. Long training must mold the mass to coherence, shape it of a piece from C.O. to trooper. The C.O. cannot get organic cooperation by improvisation from an elaboration of horsemen—be he the greatest. Cooperation follows when horses and men have trained to cavalry, to familiarity with one another, and to ultrasensitive obedience; and when reconciliation has produced unity of officers' divergent plans and contradictory procedures. And then the mass must be accustomed to sustained movement in big, complicated formations across vast, open terrain. Anything less in cooperation and obedience will produce not focused, shattering force of concentrated strength, but self-destructive collision, gaining nothing to match its losses in men and horses.

Where could the South have found time and instructors to meet those conditions and teach what had to be learned? Let's face it: even in close-order drill and compact maneuvers, as little as one division has to have been trained; it must know the evolutions. Only after practice, practice, and practice can such a unit do by second nature what the simplest moves require. Consider some in the new manual.

> The closing of gaps in the front rank by squadrons from the second and third;
> The interactive collaboration of the first two ranks, as when the second usually carries off the palm by flanking the objective, or when in an emergency the second goes by squadrons in columns through the pell-mell of the first and deploys: trying to secure what the initial thrust has seized;

And then the third rank: joining by serving as reinforcements, say, or moving sideways to turn the enemy's flank, or being the weapon in the commander's hand, easily and quickly applied in actions other than the thrusts at the start: the lash to the enemy's weakened flanks, and the ace up the sleeve in a war for keeps.

Each movement varies also as directed against cavalry, infantry, or artillery; and as conducted in the open or where cover interferes, in wooded terrain or landscape intersected by obstacles. The movements in their complications and elaborations, what with their unforeseens as well as their predictables, must be practiced and drilled, drilled and practiced, else they be neither opportune nor effective in the flash of a battle's unfolding, nor reliable and efficient with the presto and prestissimo that vow success.

American cavalry had been charged with cavalry's duties. To keep itself as fit as possible for them, during intermission it must rehearse, rehearse, rehearse. But the study, drill, and practice—ergo the performance with the finish I have just described as necessary—could not be expected of cavalry that lived in the shadow of enemy sabers. Yet despite everything against it, the Army of Northern Virginia succeeded so long against a crush of numbers because cavalry served it so well.

## 10

In closing let me say a few more words about raids. They bask in too much favor in some circles. Against green troops and armies dependent on elaborate supply nets; and in countries with few railroads, limited telegraph, and bad roads; where forests hide movement; where opposing cavalry cannot match the raider's—then and there, raids can work. Theaters of the American Civil War favored them therefore, yet they fulfilled promises but seldom then and there. They lost effect when repeated, could barely be felt. Gradually, as the armies learned countermeasures, raids could so easily be parried that they did nothing by war's end, came to be regarded as aberrations, idiotic moves.

It follows from what I have said—obviously—they would wilt as fast on our fields. Raids here would be easier to parry than over there. European raiders furthermore would find little worth destroying, what with the rail net to move supplies. Moreover, what if they flash cometlike across the fighting's firmament? The telegraph net will soon confirm existence and determine route. The raid that failed would suffer not only the loss of matériel and harm to the flesh, but also the odium of ridicule. Of course anything in Europe, productively begun and forcibly pursued, can succeed as well anywhere, given luck. But we have no place for raids in the calculated waging of a well-planned campaign. True, we have sent divisions forward, for example at the start of the campaign of 1866 and in the battles of 1871 against General Antione Chanzy and General Louis Faidherbe. True, those divisions could damage; they damaged the enemy's rear and flanks. But, unlike raiders, the divisions operated out of bases and with objectives.

Yet the equestrian spirit, the dominant trait in the German and the American, makes each so like the other! The spirit delights in bold resolve and brave deeds; it takes its reward in devotion to the corps and dedication to service. So if any branch of our forces concerned itself with military science in America, and learned something from the war, that branch had to be our cavalry. Thus, in closing, let me speak to my colleagues of that branch.[14] Stuart on his first raid felt moved to throw two bridges across the swampy Chickahominy and with his own men. I discuss the episode in the chapter on engineers. Notice it.

# 5

# Artillery

*People will look back at Charleston and say, "Modern artillery began there."*

Introductory—Field Artillery: Cannon and Guns; Organization; Tactics; a Lesson for Europeans—Siege and Defensive Artillery: State at the Time, Cannon and Guns of the South and North, Their Service, Effectiveness and Achievements.

In the peacetime armies of the United States, the branches with technical training and specialized functions got the attention, the favors, and the respect. West Point's best therefore took up artillery and engineering. In the regular army, Sherman, McClellan, and Jackson followed artillery; Lee and Beauregard, engineering, and need I say more by way of example? During the war, the best officers continued to devote themselves to them. In the North, to the observer's astonishment, the field artillery grew from seven batteries to three hundred. The fact supports my thesis that the best went to artillery, for this spontaneous volunteer artillery achieved everything that could be expected of regulars.

At that time in artillery, the North, like the rest of the civilized world, struggled with the transition from smoothbore to rifled. The questions—Which would be better? Which ought we to have?—had not been answered. We shall be looking at the American artillery in midpassage; it illustrates the transition. In the crossing from one to the other during this war, most American arsenals and foundries produced smoothbores and mainly of bronze. Only Robert P. Parrott, at the Cold Spring foundry north of New York City, cast the newcomer in field and siege, as well as producing the ammunition. We take up his methods in the section on siege.

## 2

Of smoothbores the Americans had, first, the Napoleon: the well-known, short-barreled, chambered, bronze twelve-pounder of French provenance and adopted by the United States. It compared in effect with our old twelve-pounder. I mean it equaled ours up to 1,800 paces with ball, 1,000 with fuse-detonated shot, and 400 with canister. Both sides, Union and Confederacy, preferred the Napoleon rifled.[1]

I must note the remarkable fact that the North also took out of mothballs and used a heavy ordnance of twelve- and twenty-four–pound howitzers, smoothbored, and better left in place than moved. Today's rifled artillery delivers high trajectory with small propellant. This Northern artillery had not achieved that state of the art. But it could nonetheless hit defiladed targets with superior accuracy: believe me; I saw it. Whatever the shortcomings, the North considered this artillery necessary to cope with dense forests.

The North's rifled artillery consisted mainly of ten- and twenty-pound Parrotts, that army's counterpart to our four- and six-pounders. Range of the Parrotts: over five thousand paces; propellant: one pound; ammunition: shell, canister, shrapnel, and the rarely used solid shot; barrel: cast-iron reinforced by a wrought-iron ring at the breech.[2] The twenty-pounder served as stationary, the ten as mobile. Let me note in passing that the Confederacy considered the enemy ten impractical and often expressed contempt for it. Heavy, clumsy barrels made those rifled guns hard to maneuver, and the twenties harder to manage as field pieces. The North had a few other types of rifled artillery, little used. I omit them because Parrotts replaced them after a while.

The South had the same artillery, also the Whitworths: a fifteen-pound breechloader and a six-pound muzzle-loaded cannon, both more maneuverable than the Parrotts. Especially the cavalry, half its ordnance being smoothbored, liked the six-pounder for its lightness and mobility as well as its adding a rifled to the smoothbored arsenal. Imported from Britain, they must be run through the blockade, so their numbers diminished relative to other types as the blockade tightened. Southern foundries later in the war cast Parrotts and Napoleons but only as field artillery.

Confederate generals cherished smoothbores with a passion equal to Union generals' loathing for them. In 1863 the question still impended. Its urgency prompted me to ask it of Lee. Adopt rifled altogether? "When a battle hangs fire, and fighting rages to and fro," he said, "bring up the smoothbores. Nothing beats a battery of twelve-pounders like them. Give me the smooth, roaring at four hundred to six hundred paces. I mean not only the power of shell and canister but also the demoralizing effect of thunder at close range. I know the value of rifled in the open. But at the times I just mentioned, smoothbores excel. I have decided beyond doubt to keep as much as half smooth." To appreciate Lee's preference, bear in mind that most of this war's battles took place in wooded terrain, visibility seldom more than eight hundred to one thousand paces.

In the same measure as the North enjoyed the industries to produce ordnance, so the South proved tardy in getting it. Especially in siege artillery the lack of manufacturing hurt. (I discuss siege later.) Already at Bull Run, the North, able to manufacture, could boast its first rifled cannon, twenty-eight pieces, but knew little how to profit by them. They opened on an imaginary enemy miles off, voiding the superiority of rifled cannon over the enemy's mortars. The South at Bull Run, putting a stopgap remedy to deficiencies, had set one-pound mortars on improvised carriages and rolled them into battle. As the war dragged on, the South's incredible effort brought inchmeal success at being self-sufficient: a powder factory each in Augusta and Columbia, using saltpeter from Tennessee; in Richmond, artillery foundries; in Atlanta, small-arms factories. Meanwhile the South confiscated enemy weapons to arm itself.

Both sides organized artillery six to a battery, a captain in command of 145 men, 72 horses, 2 munition wagons or caissons, and various baggage. A piece of this artillery included a carriage.[3] Six horses pulled it. Though designated by number, more and more batteries took names of well-known captains who commanded them or of famous battles. A major commanded three or four batteries. Six made a regiment under a colonel. Every division had four batteries: in the North, three rifled and one smoothbored; in the South, half and half. Each battery had its battle flag. Streamers, honoring the battery for battles fought, joined the flag. Carried by headquarters, flag and streamers served to signal the battery of a change in position. American artillery arranged itself and moved in an order similar to ours but more complicated. In a noteworthy feature, on every march and in every evolution the ammunition wagon had orders to stay with its cannon.

The numbers above show that the North had a battery for every 1,200 to 2,000 men in arms; the South, one for every 1,000. In 1864 this amounted in the North to forty volunteer and five regular regiments plus thirty organized by states for their defense and remaining under the governors' control. The South counted about half that strength.

Tactics developed little by little from simple beginnings. Early in the war, both sides attached artillery to brigades. Strong emplacements of reserve artillery, newly popular nowadays, could rarely be possible in that war because dense forests made visibility poor. Hence negligible artillery in reserve. Rifled cannon would begin the attack, trying to make the enemy deploy, firing on the reserve as well as the main force. Smoothbores meanwhile accompanied the attackers and fought at close range in dense forests. Southern tactics diverged from Northern in an important respect. At the strict and repeated order from Lee and Jackson, theirs fired at Northern artillery only when no infantry offered. Otherwise: infantry first, foremost, and always the target. Many a time Lee told me of his pains to have the order obeyed. For, over and over, the natural rivalry of counterpart arms betrayed each to disobey orders—the one to fire on the opposite, each to prove itself to the rival.

I have already pointed out that, visibility poor and terrain rugged, artillery seldom massed for attack. A few mass attacks did occur. Chancellorsville, May 2, 1863, for example. Jackson's brave corps assaulted Fairview Hill twice, success nil, losses enormous. Lee himself—how unlikely of him—took part then. He drew on artillery that happened to be passing, and on reserves, and set up a battery of twenty pieces. The well-fortified hill lay to the front at 600 paces. The improvised battery could deliver rapid, sometimes enfiladed fire. Union accounts describe the effects: savage. These helped Jackson's third charge succeed. At Gettysburg, Lee massed too: some 100 pieces to bombard Cemetery Hill in preparation for storming. At first only 45 Union answered. Gradually more assembled on the hill, numbering at last to equal his, and answered with grape and shell at 1,200 to 2,000 paces. The cannonade lasted about ten hours, many munition wagons blew up, and many limbers crumbled in a martial spectacle that meant nothing to the battle's development.

Later it spelled damage to the Confederates when they stormed the hill. Artillery, there to be used against artillery in the duel, hurled a hail of lead pernicious against flesh on that open, sloping surface. The assault failed. The spectacle also expended arsenals of ammunition. What with depots far, another engagement would be dicey: one more reason for Lee to withdraw after the one that ended without decision.

I give both sides high marks for maneuvers in battle. American officers in training and commissioned, as I have said, preferred artillery over other branches. The favor produced excellence. Infantry as early as the second year must abandon the column for the line because of artillery's skill and power. The column prevailed in Europe at the time—we have only graduated to the line after many lessons in our last war—confirming the Americans' vision. American artillery tactics, however, came to resemble ours when their war ended. Good horseflesh and better horsemanship made Southern artillery the best for agility and quickness. One author, though not favorable to the South, conceded: "If Union artillery exceeds Confederate in men in the ranks, quality of weaponry, and numbers of horses, the Confederate moves well, works efficiently, and fights without fear. Union must therefore often exert the most profound efforts lest it be outdone."[4] The mobility of Confederate ordnance meant especially the horse artillery (the cavalry's heavy weaponry) and it enjoyed regard throughout the Confederate forces.

### 3

The war began during a crisis for siege artillery: no moment more consequential in its history since the invention of gunpowder. Rifled small arms, albeit shakily, had passed their test. The technology could produce rifled field artillery; it could deliver as smooth could not; and it played its first role in war in 1859. The obvious question obtruded. Would siege guns, rifled, prove themselves in serious battle? True, in tests they shot far and hit, rarely missed, what they shot at. Theory (always the upstart), by its surmises, conjectures, and speculations, outran practice and thumbed its nose at practicality besides. Had theory established the effect rifling would have on siege? I recall Prince Hohenlohe Roerdansz's pamphlets on artillery then, and my own. Mine, written from the standpoint of an engineer, got me this assignment to America.[5]

So much for theory; practice must rule. Eagerly, nervously, intently, expectantly, tensely the world's military establishments watched the Battle of Charleston. Heavy artillery, ironclad ships, and ingenious defenses in lurid interplay took their first test. The questions about siege artillery could have been put to nobody better than the Americans. The industrial North: a people inventive, abundant of technology and money, tireless, passionate for enterprise: if anybody could solve in practice what had been posed in theory, they could. Answers came thick and fast. A multitude of consequences followed. I shall report to the extent that readers may be interested, and in aphoristic form, and as much as space permits. I shall refer mainly to Charleston because I spent time with the besieged there.

Though more interested in the offense's artillery, let us glance at the defense's. Little of recent manufacture, of course; mostly assembled from pieces out of mothballs, off decommissioned ships, and from captured forts; what a checkered medley inside the walls! Design and fabrication had improved by leaps and bounds, but this hotchpotch in no way represented that progress. Plenty of eight-inch Columbiads, eight-inch seacoast howitzers, and forty-two-, thirty-two-, and twenty-four–pounders off ships, nearly two hundred in all; but need I ask what good this gaggle of antique smoothbores against the firepower of modern technology arrayed against it? What good? Almost no good.

The South had recognized its problem soon after war broke out but lacked the means to solve it. A nasty one: mines fallen into disuse, no foundries, and machine shops able to do no more than repair ships and service the railroads. As for technicians and craftsmen in the iron-working trades, their number barely counted in the Confederate states. A problem of such magnitude would have stumped a lesser people. These—united in energy and resolute in rebellion—ran the simpler tools through the blockade, erected machine shops, set up work places, trained workmen, built cannon foundries, installed steam engines, and reopened the long-neglected mines. Richmond became the seat of most cannon and small-arms manufacture. There the Tredegar Works enlarged the plant, expanded to a concentration of arsenals, and consolidated with a big cannon foundry that cast ten-inch Columbiads and seven-inch rifled Brooke guns in addition to field artillery.[6]

In artillery the South could muster nothing to touch the battle effectiveness of the North's except those just mentioned: two big, cast-iron blasters. The huge Columbiad, besides being accurate enough up to two thousand paces, had the advantages of smoothbores: inexpensive, easy to build, simple to maintain. The Brooke proved unreliable. It answered the needs of range and accuracy; but though three wrought-iron rings reinforced the seven-inch rifled bore at the breech, the breech cracked too often despite them. A breech took some replacing. In the end, most of the smaller rifled cannon, muzzleloaders, suffered the same fate. You would see one after another in position but disabled, broken breech cast aside. Worse, watching a stationary battery (Cheves) open fire from behind a screen of forest, I saw two of eight carriages able to maintain an elevation of twenty-five degrees.[7] Being naval apparatus, they bounced like crazy furthermore, leaped and bounded so that after a few shots they could be fired no more. Have I not shown by now a defensive artillery able to meet the most critical needs in a pinch but not fit to fight a war?

In August 1863, two Blakely guns arrived. A ship ran them, ammunition for them, and nothing else through the blockade. It could run nothing else because the Liverpool-built, twelve-inch, rifled, cast-iron monsters made a cargo themselves. Alloys had been added to harden and toughen the barrel only at the breech. In addition, three heavy, wrought-iron rings reinforced it. Triumphantly the Confederates mounted the Blakelys at White Point Battery, part of the bastion of Charleston's defenses. At the first shot, the breech of one failed, the barrel burst. Alloying had not strengthened the breech but weakened the barrel.

The Confederates had remained ignorant of the range and accuracy of the newest ordnance. The Federals, after devising some of it, manufactured it with a vengeance. We should not be surprised that the first shots of the bombardment of Fort Sumter (1863) astonished the Confederates, hitting them bull's-eye when they thought themselves out of range at five to six thousand paces. Being hit worsened the fact of helplessness against what fired at them. Trying to harass the attacking batteries, they managed a few rounds from James Island. Those pitiful efforts did nothing but lose another gun. The Confederates could assert nil to protect themselves. Their defense must be to hold their fire, save their ammunition, and keep their powder dry until the last stages of the siege.

Meanwhile, interest declined in the artillery of the defense. Every observer had overestimated its strength. The less the interest there, the more the eyes of Europe turned to the North's progress in rifled cannon. Indeed, testing began at Charleston. Anyone interested in learning more about it should consult the little work by Jacobi.[8]

I have mentioned the defense's standbys: the 42-, 32-, and 24-pounders. Newer included, of smoothbores, first, Columbiads by the Rodman method: iron cast around a core chilled by a stream of cold water while the casting's outside stayed red-hot until cooled from the inside. The method improved the casting because the iron did not crystallize but hardened in stages, creating uniform texture and reducing molecular stress. Bores needed such improvement because greater pressure of explosion led to progressive cracking from the inside out. By this, the Rodman method, the bore hardened more than the jacket, diminishing that problem. It proved to be the best method. The three sizes of Columbiads thus cast (8-, 10-, and 15-inch) fired a 64-pound projectile with 10 pounds of propellant, 126 with 15, and 400 with 35, respectively. The 15 could fire a heavier one, 430 maximum.

Modern times make stern demands on big guns. We have seen in Charleston's defenses the 8-inch that did not measure up. The mobile, serviceable 10 won favor, commanding respect at distances as great as 2,000 paces yet no less effective against fortifications at close range. In tests, one showed no deterioration of bore after 1,400 rounds using 15 to 30 pounds of propellant. The 15, though not as accurate, had an advantage over the smaller: the power to smash the hulls of ships, invaluable for short-range uses such as harbor and river defense. Especially effective against ironclads, it knocked out the *Atlanta* and the *Tennessee,* for example, each with a few hits. The first to strike the *Atlanta*—400-pound projectile, 35 pounds of propellant, 400 paces—ripped a hole five to six feet long through layers of iron and wood, and disabled forty-eight men. The second, third, and fourth wrecked the steering. She had to surrender. In the Battle of Mobile Bay the 15 proved even more its mastery of ironclads. A hundred hits by 9- and 10-inchers had not broken the armor of the ram *Tennessee.* Two from the 15 smashed through. They stopped in the wood beneath but hurled splinters and fragments to wound much of the crew, forcing her to end her heroic resistance.

Americans, urgent of need and enormous of demand for artillery, treasured another Columbiad advantage: ease of maintenance and at low cost.

The Columbiads had their drawbacks, though, the worst being bulk. They illustrated the word *massive*. The barrel weighed over 4,700 pounds. Witness the dimensions, the enormity of the colossus, 190 inches long and over 48 at maximum diameter. Long and fat, therefore heavy and clumsy, they found little welcome on ships. Most of the navy preferred the 10-inch Dahlgren gun. Short range also spoke against the Columbiad. Taken together, the disadvantages relegated Columbiads mainly to coast defense.

The Dahlgren, the second smoothbore, though of cast-iron like the Columbiad, looked different because of the abrupt taper, contrasting with the conical Columbiad. Essentially they differed in manufacture. Founders cast the Dahlgren as a cylinder, then bored and tapered it. Cylindrical casting supposedly promoted cooling everywhere equal.[9] Against ironclads the Dahlgren could not equal the 15-inch Columbiad. The 9-, 10-, and 11-inch Dahlgrens came about up to Columbiads in those sizes.

Parrott guns predominated among the rifled artillery: muzzle-loaders cast by the Rodman method. They saw their greatest use against Sumter. A wrought-iron ring strengthened the breech. Half as thick as the bore, it extended slightly forward of the point of the projectile in the chamber. The manufacturer applied it like that of the Confederate Brooke. Heated red-hot, then cooled in water, it shrunk fast to the breech. We have already met the well-known Parrott field artillery, also represented at Charleston. Other Parrotts included the 100-, 200-, and 300-pounders; respectively of 6.4-, 8-, and 10-inch bore; 80-, 150-, and 200-pound projectiles; fired by 10, 16, and 25 pounds of propellant. Range, 5,000 paces; accuracy, 60 percent; projectile cylindrical ending in a frontal cone. The projectile's leaden jacket had nothing to do with putting on the spin. The jacket would protect the rifling. The spin came off the copper ring, around and a little bigger than the projectile, and traveling in the rifling. The shortened spiral near the muzzle accelerated the spin.

The Parrott gun, popular in sieges and the navy, exceptionally accurate, had been a favorite in the Union army, especially for hitting its targets. Several exploded at Charleston, however, and the six before Fort Fisher that killed or wounded forty-eight. The mishap so depressed morale that cannoneers could scarcely be brought near the Parrott. At a stroke it lost its good name.

Indeed, both sides learned of this rifled piece that burst again and again. The evil usually occurred at the back, around the reinforcing ring at the breech. Investigations determined part of the cause: the missile would explode in the barrel. Careless manufacture in turn probably abetted that explosion and the one when the missile left the muzzle. Parrott himself—he ran the West Point foundry—laid it to Morris Island's sand. Any hint of breeze raised those bits of quartz in clouds so thick as to foul the bore, muzzle to breech. After his diagnosis the cannoneers must swab it clean with the slush brush, not every twenty rounds—every three or four. Again, a casting fault may have caused the explosions. I would blame something else. The copper ring at the missile's base to give it the spin permitted a vibration of the missile to and fro despite the leaden jacket. Finally we must notice people who saw the fuse as culprit. The Parrott could fire missiles that hit the target from overhead, and some went off ahead of time.

We have added up defects and problems. Despite them the Americans must get the credit for pioneering in heavy rifled ordnance. We have not said, but should, that Rodman introduced prismatic or coarse powder for it. Yes, European elaboration of this artillery has surpassed everything of those times there—let that fact not be missed. But we Europeans have excelled because of what the Americans did.

What of the service of this artillery? It meant strain and vexation for battery commanders. We have discussed artillery's many difficulties, annoyances, and problems. Only the closest attention and strictest care could deal and cope with and in a measure alleviate them. Crews must therefore be alert and at work, and battery commanders accordingly present and vigilant, whenever guns fired. They could deliver once every five minutes with the hundred-pounder; every seven or eight with the two hundred. Extra transport crews for ammunition had to be assigned at Charleston: not merely to move such projectiles, but also because of distances from magazine to battery and deep sand in between.

In the Confederate forces on the defensive a piece had but one change of crew. Under fire, either it would be withdrawn and the relief take over, or it stood unmanned until it could be served again.

With quarters nearby, however, enough men could service even in the fiercest fighting. Artillery of the collateral forces on James Island had no reliefs. There the artillery of Beauregard, Bee, and others kept crews always available. A sentry posted for each battery alerted the crew when needed. The offensive demanded twice as many reliefs because troops quartered farther from the action. In the bombardment of Sumter, crews of the siege artillery's breeching batteries alternated four hours on, eight off; other artillery, twelve on, twenty-four off.

The Confederate's defensive artillery could achieve no spectacular results, as I have said. Defensive batteries therefore limited fire to critical moments. It counted then. Fort Wagner and Union monitors traded blows once. I counted fifty-one hits on monitors: a clang reported each. Cleared for action a monitor presents a small target. Fifty-one by six pieces in three hours accordingly deserves a good score. Wagner's artillery, again, aided by two mortars doing their part, defended against sappers from the fifth parallel—so well that two and a half days' work did not advance the assault thirty paces. According to attackers' reports, sapping would no sooner begin than the defenders' fire render the complex of trenches unsafe. Sappers at last grew cross and sullen. This result exceeded by far what the defenders expected of their artillery.

Of course the defensive could not equal the offensive even for interest. It told against Sumter's walls—nearly demolished the fort—and at ranges never contemplated before, provoking astonishment. Whatever the doubts about heavy artillery's capabilities, those successes scattered them perforce and redirected the sciences of artillery and fortification far away in Europe.

Sumter stood on a level sandbank in the middle of the harbor. The foundation of square sandstones continued in a brick wall of two stages. The rear being toward the attack, a wall of sandbags had to be improvised to close the gorge. (See this point in my discussion of engineering.) I took post on James Island opposite the gorge. Able through my glass to count every stone and each brick, I missed nothing of the bombardment.

Hits—two per three shots, on average—smashed gaps in the walls. Only the stack of sandbags withstood breeching. But then, with bags at five to six thousand paces, breeching could not mean what it usually meant. The bombardment damaged the gorge but even more the back of the casements at the fort's front. Eight days of it battered them to hell. Nothing remained to shield the defenders but piles of rubble. By August 25, one ten-inch Columbiad remained. The rest had begun the defense but been silenced. From time to time it roared defiance. "We haven't lost our nerve. We've got the guts to stand up to you."

Walls eight feet thick protected the powder deep inside. Great store being set by those magazines, a cloak of freestone shielded them in turn, twelve thick. After a few days under fire, despite the twenty feet, the magazines showed such harm they must be cleared in the next few nights. By the end of August, however, the enemy fired as accurately at night. His calcium light illuminated the target as bright as day at three thousand paces.

Yet artillery alone could not have driven the defenders out. Sumter at the end looked more like a pile of junk than a bastion; but they took cover in fortuitous nooks and crannies and behind serendipitous heaps of rubble. An attack with boats on September 8 also failed. The attackers managed a partial landing but had to give up on their coup when the defenders, under the brave Colonel A. Rhett, let go with a hail of stones and bullets.[10]

Proponents of masonry-walled forts may therefore adduce Sumter as proof that brave defenders can persevere behind masonry even in shambles. But good sense says not to put them where its bits and pieces never stop flying about, wounding the defenders and imperiling their moves behind the wreckage of walls. Masonry failed against rifled artillery and heavy smoothbores.

Charleston provides an example in contrast. Fort Wagner, hit often and hard, held up like a redoubt, despite every effort to hurt it. Its breastworks of sand—beams of wood heaped with the fine-grained quartz—by and large protected the inside as well after as before, and kept injury to personnel to a minimum. The cannon, pulled in a little, also escaped much harm. The holocaust of August 17 knocked out but one. The rest could stay active enough to interdict the enemy until the day before the final onslaught. Wagner offered a lesson to military engineering, and military engineering learned it.

The bombs of mortars and howitzers did little to Wagner: insignificant damage. Gilmore blamed the ranges: too long. I believe it due to care in providing bombproof quarters. A few defenders stood watch and returned fire during bombardments; the others rested, safe. Besides, bombs came in slowly; the few men on duty had time to spring the short way to cover and safety.

At all events this entry in military history remains memorable for the development of heavy weapons. People will look back at Charleston and say, "Modern artillery began there."

# 6

# ENGINEERING

*In engineering, as in artillery, modern advances occurred only in theory until this war.*

Introduction—Earthworks—Masonry Fortifications—Iron as Armor—Abatises and Entanglements—Underwater Defenses—Siege Works—What the War Meant in Europe—Pioneering and the Lessons for Us, with Examples: A River Crossed by Cavalry; the Petersburg Mine Assault.

The student of this war sees what might happen in any war, given the same factors. The North had an industry and the shops to work iron; whereas the South lacked technology, a deficiency of the worst sort: two factors the student should not lose sight of, for they mattered in all aspects of this war. In addition, the character of the people will determine how a nation will fight. In this war, whenever energy and effort, the creation of better material, and persistence and the will to advance mattered, the student sees the decision go to the North. When courage, initiative in strategy, ingenuity in tactics, and derring-do with cavalry mattered, the South carried off the palm. Thus the student, by isolating factors that mattered, distinguishes the true North and the classic South. In the West, however, differences and distinctions shaded into one another. The martial initiative and the military genius of one side about equaled those of the other. There, instead, a factor of geography tipped the scales to the North. This Mississippi system let its gunboats and big guns into every part of the region. It followed that the North prevailed. The North by nature, given that access, would take the upper hand in ships and heavy artillery and join them in combined operations.

The student might therefore think the North would have the better engineers. The North did not produce better engineering; it provoked it. Because of what the North did, and out of the fear of what it could do, the South took the prize in engineering. I mean, for example, that as rifled cannon empowered artillery to astound everyone with how far it could shoot and what walls it could knock down, so the improvement must influence fortification. We must affirm what cannot be denied: had the North found itself on the defensive, we would see the greater invention and the better implementation of engineering there. But the North had the artillery; the South must defend itself from monstrous missiles, and invent and implement to protect itself. The South excelled accordingly in engineering. Indeed, the excellence added one more feature that made this war so much more than a localized, timebound rebellion, another reason to count the American Civil War one of the world's historic wars. For, in engineering, as in artillery, modern advances occurred only in theory until this war.

## 2

Let us look south to see how the Confederates built. (My personal experience does not extend to the West. I rely on reports of Donelson and Henry in Tennessee, Jackson and St. Philip on the Mississippi south of New Orleans, and most of the other defensive sites along the river.) To begin with, Southern excellence in engineering neither guided every engineer nor prepared every fortified post. In the East I inspected practically all for myself, including the ones around Richmond. (I would, for I am an engineer, after and above all.) Those and the ones in the West, what we Germans should call regular installations and permanent facilities, resembled what we reserve to the field: flimsy, meager, temporary structures. The Southerner, hating pick and shovel in his own hands, had put the Negroes to work on fortifications. The work of a Negro amounts scarcely to a third of the work of a white in value or amount. In consequence of that criminal neglect, the Union overran and captured the forts in the West.

Yet I do not contradict what I said in the introduction to this chapter. We find on the Confederate side much that compelled imitation by engineers everywhere, and many a thing that invites contemplation even now. Take the forts and batteries that guarded the mouths of the Ashley and Cape Fear Rivers at Charleston and Wilmington. In those strong, functional, purposeful bastions, a breastwork twenty-four to thirty feet thick enclosed everything and asserted the wall against the outside. Traverses, another ten, protected one or at most two cannon each, putting them, as it were, indoors. Wood in boards and planks covered the escarpment or inner surfaces not exposed to direct hits. Trenches behind the cannon accommodated crews on duty, while shelters and spaces of every sort gave refuge even during brief pauses and secured ammunition and its accessories. Plenty of inexpensive armor (earth), together with prodigal dimensions, provided room for exercise and drill.

The stalwart breastworks over extended ground encouraged freedom. The engineers need not obey prescriptions on flanking; in layout they could emphasize strength to the front and space within. The strength of wall would bear the brunt of attacks head-on and from afar and with ample room, leave flanking doubtful. (The attacker would think twice before trying to flank.) Cannon behind the wall could aim in any direction, further inhibiting flanking. Not only had they room for it, nor merely space to wheel around the traverses, but also ample ports in angular cuts in breastwork. Furthermore, having made the primary trenches stormproof, the engineers believed they could dispense with costly flanking trenches at lower levels. In some places, Fort Caswell near Wilmington for instance, they even filled in the side works subject to direct fire. Always—for good measure, whatever the circumstances—they provided within and withal the space to draw up reserves and move them against the dangerous attack. In short the engineers could forget concerns about flanking and ignore the pedantry of systems that obey mathematics.[1] Instead they could adapt to terrain, tactics, and what might result from them. Hence the simple layout: strength to the front, space inside, and flanking played down.

Earthworks meant strength; they withstood the heaviest shelling. Fort Wagner, asserting quartziferous dune sand, best illustrates this strength. The bombardment began on August 17, 1863. The ironclad *Ironsides*, six monitors, six gunboats, and every land battery in range—and with the largest calibers—tested whether the ramparts could take the pounding. Six hours nonstop they pummeled the fort. What a spectacle! Missiles, especially the sockdologers of the fifteen-inch Columbiads, blasted pits in the ramparts. Many a time the fort seemed to evaporate in its own dust. Yet little harm came to the defenses. Cannon had been pulled in, only one disabled. The rest saw the flotilla off. The second barrage, on September 5 and 6, would prepare that attack of September 7. Seventeen heavy mortars and other, lighter ones, as well as thirteen heavy cannon, and by calcium light at night, shelled forty-two hours nonstop. This time, the Union report said, the fort lay quiescent like the dunes it so resembled. The lines that had defined it had been erased; the surfaces that had distinguished it, wrecked. It looked like a random pile, regarding us across the works of sappers.

It could still be defended, but sapping had nearly reached the counterscarp. Soon, from there, sappers would tell the defenders: No escape, no retreat. The defenders must ask not if, but when they should give it up. They did not wait for the attack. They stole away the night of September 6/7 by boat and evacuated Battery Gregg at the same time.

As I have said, bomb shelters in such fortifications more than withstood attacks. Safety in them must be absolute nonetheless; they must be used for nothing but places of safety; they must be redoubts only in the sense of refuge. In Confederate works they resembled ours of today's Europe. At least ten feet of earth covered them. Thus secured, they stiffened the defense, made it tenacious. They reduced especially the nervous fatigue that must enervate in a protracted siege. I think of Sevastopol.[2]

Over a thousand men could take cover in the one underground at Fort Wagner. Long they struggled with foul air, however, a cause of malaria and yellow fever. At last somebody determined the reason. The sand of the floor absorbed and released miasmas. Take out a layer of sand, somebody said, and put a generous new one in its place. The air clarified. The effect proved as good as ventilation.

### 3

I turn to large masonry forts. Sumter and Pulaski rose in two tiers of masonry above the water. They offer European engineering no lessons except they demonstrate yet again masonry's worthlessness. I have said how rifled cannon (seven-, eight-, and nine-pounders) wreaked devastation at five and six thousand paces: at Sumter the bombardment threatened the magazine behind a shield twenty feet thick. Worse, even the missiles of smoothbores, fifteen-inch Columbiads, ripped caverns in masses of wall. With interest I observed the difference between masonry and sandbags. The fort's rear being toward the attack, bags on the quay shielded the gorge. Each held about two bushels; the stack barely reached two-thirds the fort's height; yet they kept missiles out of the thin, sand-filled casemates. The debris also offered the defenders cover. Happily they beat off the attempt to storm from boats.

### 4

Engineers applied armor (iron) to forts in pale imitation of the cladding of ships. The slanted wall could be erected with ease, so the South liked it. It consisted of two-inch plates six feet long and twelve inches wide, a double layer; and of railroad rails interlapped, a double layer; the whole tilted thirty degrees. It would, in theory, like naval armament let no missile hit straight, only askew, and rob momentum and brush it off in ricochet. Compare in the naval chapter the design of the *Merrimac*. Iron in land fortification unfortunately never got beyond theory, never to prove its mettle against serious attack.

Let me mention an unusual incident in the armoring (with rails) of Fort Caswell below Wilmington. The stonework had been done; the glacis raised and topped with traverses to protect the escarpments from indirect fire; and as I have mentioned, the protected trench heaped with earth. This being a fort with casemates, the time had come to shield its platform. Only the breastwork, though twelve feet thick and adequate in times gone by, would not do against the wallop of modern artillery, nor would the rampart take a thicker one. The engineer, Lieutenant Colonel Taylor, decided to face it with rails joined by clamps in a double stack to be covered with six to eight inches of earth. (They might use earth in the breastworks of any fort then.) The builder, knowing nothing of the stoutness of rails, put up a little wall on trial. This fragment he tested with the biggest gun around and at the shortest range. He built and shot until it could stand the worst and show little harm. Thereafter, when fortifying, the engineers included earth—always.

### 5

American military science liked abatis and entanglements, especially the natural ones erected easily of trees in a country of forests. Such barriers on James Island proved solid and impenetrable but flawed because too high to be commanded from the fort. The popularity extended to field fortifications, and the flaw worsened there.

At Chancellorsville the one of trees to protect the Union position, involving three hundred to five hundred paces yet commanded from trenches, neither asserted the breastworks the Federals wanted nor posed the problem the Confederates hated. The Federals erected it to impede the Confederates. The Confederates advanced, shielded by it.

## 6

Everyone knows the sensation two defensive innovations caused. I mean torpedoes and underwater obstacles.[3] A lot has been written on them; what has been written has been assertive; and the writings have been much read. We must listen with care to the noise, like everything about this war. Torpedoes did explode against their targets, especially in the West, notably in the Battle of Mobile Bay against Admiral Farragut's incursion. But intimations by Southerners led me to believe what caused some of the statements: not reality but fear. Therefore I shall keep it simple and true by sticking to what I learned at Charleston.

The repeated tries to close the harbor failed, every last one. Either a torpedo had been laid and anchored to float, or weighted to ride below the surface and rise and fall with the tides. The first could be seen and gone around. The second soon went to the bottom; if not mollusks and other crustaceans, then seaweed dragged them down. Of the many laid off Fort Wagner an accident turned most into land mines. A northwester ripped them loose and strewed them the length of the shore between Wagner and Fort Johnson. With caution the many on the beach of James Island had been disarmed, at Wagner a few perhaps removed and buried.

A netting supposedly entangled the screw of a monitor. My inquiry brought a knowing smile. Such a device, I have reason to believe, existed only in the imagination of the captain who reported it. Experience had taught what would be effective: forget reality, rouse fear. Exaggerate, embroider, embellish the underwater devices in word if not in fact, and terrify people of what they could not see.

Other things tried during my stay in Charleston included floating rockets against the attacking fleet on the high seas, as well as an interesting submarine torpedo boat. It operated well in the harbor but failed a number of times in open water. Out there it did manage to approach the frigate *Ironsides* with a torpedo but caused little harm.[4]

The torpedo boat probably sank on the second foray, costing the brave captain his life.

## 7

A sketch of an assault must precede an opinion of siege operations and the work of engineers and the pioneer corps. Quincy A. Gillmore, general of engineers, commanded the assault on Fort Wagner. He landed his 11,000 on Folly Island. Would he show why fate had put that name on that piece of ground? On July 10, 1863, 3,500 addressed Morris Island from the south: surprise attack. Masked batteries supported it. Field artillery alone opposed it. Emboldened by success, the attackers continued so soon as the next day, trying a powerful surprise on Wagner. Failure. Defenses had been underestimated, the attack prepared merely by shelling from four monitors the prior night, and the result as dismal as the success had been brilliant. Hence the decision to get established on the island before attacking again. Construction lasted until July 18 on the emplacement of 4 batteries, including 1 of mortars. In all, 27 cannon and 10 heavy mortars glared at the objective. A parallel (so the Americans call it) 1,600 paces from the fort served in particular to shield them.

From this base on July 18 a second attack opened on the beleaguered works. The storming party, 4,500 men in column, left at dusk because the second line of defense ought not be able to see them in twilight. But Fort Johnson signaled *here they come!* Every outwork of Charleston opened fire. The attackers, Negro regiments among them, proceeded with grit and a will. They scaled the breastworks several times, even won part of the southeastern side for a few hours. Darkness and sustained fire from every corner forced them to give it up. They suffered heavy casualties besides. The temporary success and its main achievement had come to grief on the unhappy choice of dusk. Fighting in the dark, friend often shot friend.

That part had failed, that vehement, self-ruinous assault. But the pitched battle had begun. The Federals chose nothing less than a frontal on Wagner and the destruction of Sumter at the same time. First they looked to their own security, dreading that it might not suffice. They armed the outlying parallel, the first line of defense, with field artillery and Requa batteries (a kind of machine gun able to swivel and traverse).[5] Distance too great for safety lay between the parallel and the objective, so they put an oblique picket there.

It ran across the island and bent into a concave at the flanks, thus also defending the beach at low tide. A swamp to the left offered the enemy a chance at that flank, so a defense even there: a wire entanglement to the start of the picket at Lighthouse Creek.

This work lasted until July 24, a long labor. More began at once on a second line. Measures went into it like those into the first and later the third and fourth: measure upon measure to secure against sorties. Work ended, and the attack could begin, with the siting of eight batteries, each of seventeen Parrott guns. They, together with artillery already there, would bombard the objectives, primarily Sumter. One piece constituted a minor but instructive experiment in artillery and its emplacement, the so-called Swamp Angel, situated in the middle of a swamp, a ten-pound Parrott behind a breastwork of sandbags on a foundation of fascines.[6] Put there to hit Charleston, and firing thirty-six times, it did hurl perhaps twelve missiles containing Greek fire into the place. None exploded. The overtaxed barrel gave up on the thirty-sixth; the gun itself exploded.

On August 16, breeching batteries (twenty-three pieces) opened fire on Sumter.[7] Sapping meanwhile approached Wagner. The nearer it got the more the Confederate superiority in small arms counted. Therefore the trenches advanced slower and slower. In the final stages they gained but thirty feet in two and half days, working day and night, as enemy rifles picked off sappers. To suppress this sniping, the Federals set up batteries of light Coehorn mortars in the third, fourth, and fifth parallels and showered the fort with their bombshells. The defenders evacuated, as I have said, but not until the brunt of the attack reached the counterscarp itself.

The attentive reader has spotted the blunders. Lacking understanding of the situation, the attackers wavered between extremes: rash one moment, fainthearted the next. In earnest but with almost no preparation they hit the stout Wagner twice despite the narrow front so unfavorable to attack. They ought to have seized the chance to consolidate a position as near the fort as they could and made things as tough for it as possible from there. Then, 1,600 paces off and with anxious caution they began trenches and parallels as if they feared a sortie at any moment. Given Wagner's narrow, almost unassailable front, they ought to have been bold, pushed ahead, opened the first parallel without hesitation at 800 paces, the second and third as near as possible behind it, and from then tried to storm the fort on

July 7 at dawn. Being the first, and intended by concentration and force as the only attack, it would have been the likelier to succeed. It would have gathered nearly everything into Northern hands, saved time, and spared sacrifices.

With interest we note nonetheless how the Northerner understood technology: its uses on the battlefield both in the attack and behind the lines. Back there, steam ferries, observation posts, lighthouses, good roads, and other amenities enhanced traffic, aided communication, facilitated service, and strengthened support.

And the defense? I sided with the defense, but I cannot blink at the facts. The defense did not measure up. It fell short of both the enemy's challenge and its own circumstances. True, in that swamp-island and in that unhealthy climate the garrison must have the daily dose of quinine. Of course it made them look sick and feel lethargic. Not even alert, how could they be nimble and quick against the forcible attack? Again, the terrain promised success to sortie, counterattack, and a hundred small defensive operations on land and water. Yet the defenders availed themselves of none. Worse, the enemy reconnoitered nearby, brazenly close, and the defenders permitted it, did not strike at his rear or do anything else to make the probes ruinous to the prober. Let me emphasize the reasons: (1) apathy exacerbated by climate and quinine; and (2) loathing for work with the shovel.

## 8

Did European and especially German engineering learn from what the Americans did? Easy question, obvious answer. The Americans built with what the land offered in abundance and cheap: earth. By "width of wall" the Americans meant thickness of earth there, but also how much on hand there to replace that blown from bombproof shelters by enemy bombardment, besides enough (as I have suggested) to extend the breastworks it needed in a new direction. Breastworks at Lighthouse Creek lengthened and shifted as the batteries turned by degrees at the rate and to the extent of the enemy's flanking them to the left. Europeans now understand accordingly the advantage to storing earth in breastworks and that the gain increases in positions short on lateral space to pile the excess and to extend and reinforce traverses. French reports mention ample earth, extension, shifting, and similar measures in the defenses at Belfort.[8] Indeed, the American experience not only influenced how earth be used, but also that engineers prefer nothing else and want no more masonry in America or Europe after the American Civil War. We over here also began to put traverses in the parapets of earthworks, a feature considered in purpose and scrupulous in design. To those fortifications we added the bombproof shelters needed, and built them suitably stout. Gradually we, too, resolved not to do more to defend a place than put up impregnable walls, stay behind them, and not try to defend beyond them.

I have just said that the Americans gave up defending anything beyond the main walls and that we followed the example. We, however, continued to believe the defense of outlying flanks essential. Again, we have fixed the breastwork width at six meters; they believed it had to be eight and more, protection to spare. On our side of the Atlantic as well, notable efforts have succeeded with iron as armor. In America it served where nothing else would do in limited space, and in shipbuilding for the duration. Otherwise the Americans chose earth.

## 9

In the war's last period, pioneering expanded and transformed. What did we see among the infantry—and in our chapter on tactics—in the battles of 1864 and at last at Petersburg? Less gun and bayonet, more axe and shovel.[9] The right men and plenty of material furthered pioneering, of course. I mean axe-savvy farmers of skill and experience, and Nature's bounty of forests, as many and as much and as apt as nowhere and never and nothing before.

American pioneers, though not as well trained as ours, built bridges as if born to it. Even the cavalry never lacked axes nor hatchets nor men who knew how to hew from forests the stuff of limited, temporary fortifications as well as the skin and bones of bridges small and big. Let Von Borcke relate the example of Stuart's White House raid in the summer of 1862.[10]

"About ten o'clock we had an hour's rest to feed our horses, and then rode on all the night through towards the Chickahominy River, which we reached at five o'clock in the morning. From the reports we had received we expected to find little difficulty in fording the stream, but who can describe our astonishment at finding it so swollen by the rain which had fallen in the mountains during the past twenty-four hours that the water was more than fifteen feet deep! At the same time our rear guard announced that a whole division of enemy was our track.

"Every one felt the weight of the danger that threatened us, every one looked with anxiety towards our leader, of whom once again I have occasion to speak with the warmest admiration. With the greatest possible calmness and coolness he gave his orders and made his arrangements.

"Two regiments and the two pieces of horse-artillery were ordered, in case of an attack, to cover our retreat; whilst all the other available men were dismounted, some of them being employed to build bridges, the others to swim the river with the horses.

"A bridge for foot passengers was hastily constructed across the stream, which was about ninety feet in breadth, and the saddles, &c., were carried over it.

"All the swimmers took the unsaddled horses through the river, some riding them, others swimming by the side, with one hand holding the mane and the other directing the horse. This last expedient I thought the best, and in this manner I took sixty-five horses myself through the angry torrent.

"After about four hours' work a second bridge for the artillery was completed, and more than half the horses had reached the other side of the river; also the prisoners, about five hundred in number, and hundreds of captured horses and mules. Then the first cannon was drawn by the soldiers across the bridge, which stood the test well, and soon the second followed, and then the reserve regiments.

"Towards noon all were in safety on the other bank. General Stuart was the last man to cross the bridge, which we then destroyed."

Pioneers mean as much to the army as do the combat forces. We Germans can make something of our pioneers by multiplying the corps itself, or by proliferating and propagating those attached to the foot soldiers. But we cannot create them in the cavalry; short-term enlistments forbid creating them there or in any other combat arm. Ought not the German cavalry assign units their own pioneers and mounted like the rest of the troopers, then? I pose the question, out of the American experience, for our consideration.

Let me end the discussion of pioneers with an episode of interest: the Petersburg mine assault. Union forces tried with an enormous subterranean explosive to cleave Confederate fortifications there. On

June 25, 1864, action began underground: a tunnel 522 feet long and with laterals. On July 23, digging ended. On July 27, the explosive had been placed and armed. Despite official secrecy, rumor spread among friend and foe. Beauregard, thus able to approximate where the tunnel would open, had his batteries accordingly placed. He took care to keep these preparations from the enemy, as well as the fact of concentrating his thin, weak forces at the spot. Thin? Weak? At one man for every seven feet, Confederate lines ought be called porous and frail.

On July 30, the mine ready to explode, General James H. Ledlie's division prepared to storm the breech the mine would open. General Orlando B. Willcox's, and General Joseph H. Potter's divisions, along with General Edward Ferrero's Negro troops, formed in support. At 3:30 A.M., fuse lit, everyone awaited the blast on tenterhooks. Everything hung fire.

They waited.

Nothing happened.

It might have been the old story, the opera bouffe of great expectations and a letdown, the rise to climax that proves to be a fall to anticlimax. Every man held his breath.

Nothing.

Someone discovered that the fuse had gone out. Someone else touched flame to it again. At 4:42 a cone of fire erupted. Two cannon and their crews flew into the air on a concussion of black powder. A crater 60 by 240 opened. The blast deafened and shook the defenders. One, General Stephen Elliott, a veteran of the Sumter defense, somehow rallied his men, sprang to the breastworks at their head, took a bullet, and fell. His men began to withdraw toward Petersburg, terrified.

The jolt had alerted Lee and Beauregard. Arriving on the run at the moment Elliott went down, they sent General William Mahone's division to the spot. Union artillery had been firing to prepare the assault. Lee and Beauregard ordered Confederate batteries to reply in force.

Ledlie headed his division into the storm, followed by Potter and Willcox. Disarray, the consequence of being disorganized, weakened the attack. Disorder indeed had softened the starch the troops had brought with them. No phalanxes from the tunnel therefore, only a dribble of disjointed units into the crater. Then the Confederate lines dissolved. Merely by proceeding, the Federals could have flanked them and mangled any resistance. Instead, used to fighting from cover, they cowered and hunkered into the crater's womb. Soon, jumbled out of their units and their wits, troops filled the crater with their bodies: a mob, a mass—disorganized, disjointed, discombobulated. Potter tried to sort out his men in vain. Meanwhile some blazed away from behind the crater's lip.

Confederate batteries opened up on it and the approaches. Lee, as calm and unshakable as ever, had by cold-bloodedness bucked up the scared defenders and kept order until Mahone's division arrived to reinforce them.

Green troops, be they black or white, take time to yellow. Ferraro's Negroes, sent up then, blustered into the crater, to no avail. The murderous barrage turned reckless courage into a panic. Confusion swelled in the craterful of rabble. The job of calming grew more and more complex, commanders being of the same rank, each in charge of his own, no one in charge of all. The aim of regrouping seemed more and more doubtful. Mahone elaborated the bombardment. Only a few in the crater's roil had the gumption to shoot back. The rest cowered behind piles of earth or braved the fire enough to turn tail and run.

What carnage among the imploded mass!

Withdraw? Reinforce? Either would have doubled the slaughter. To aid the unfortunates, attacks opened elsewhere. Confederate defenders, proving themselves brave everywhere, would not be pushed back. The Federals must sacrifice the troops in the crater. Mahone's division stormed it and killed, wounded, or captured five thousand. The Federals had come with hopes and expectations and enterprise for this mine assault. It ended in a crushing, shameful, bitter defeat.

# MEDICAL SERVICE

*The American achievement cannot be equaled; we cannot work the American miracle.*

### Introduction—Of the South—The North—Compared with Germany.

This war teaches lesson after lesson after lesson. We have learned many on the martial side. But studying men at arms and how they fought will not teach all the lessons if we ignore health and medicine. For, in them, the Americans excelled. We in Europe, emulating America, have organized the medical corps, elaborated medical transportation, incorporated the details of medical buildings, even fitted out hospitals after the American example. Our system and every wounded European owe a debt and perhaps life itself to those who worked so hard and did so much, those ingenious humanitarians on the other side of the Atlantic. Yet the example remains superior; the American achievement cannot be equaled; we cannot work the American miracle.

I must speak in general about treatment of the sick and wounded. I cannot go into detail—I have not the pages. I limit my remarks to what must concern any commander. To my colleagues of the medical corps: Read the many interesting studies devoted to the subject.

### 2

America, broadly speaking, consisted of two regions, the two nations at war: the South with its poor, scattered, isolated settlements; and the North with its cities and towns. Two systems for treating casualties followed: distribution, dispersion, a quartering on the populace in the South; concentration, a gathering into stations, camps, and hospitals in the North.[1] We shall learn of more of them, along with a word on food and diet.

In the field the South organized around the regiment and the brigade. The regiment had its chief of medicine (a surgeon with major's rank) and staff enough to provide a trained surgeon for every 200 or 250 men; a medical orderly or attendant assigned to him; and stretcher bearers, two for every hundred men. Stretcher bearer meant somebody special: strong, reliable, picked for hard, dangerous work in protracted battles. Instead of weapons he and another took into combat the implements of *their* work: one carried the poles of the stretcher's frame rolled into the canvas, the other a waterproof pouch or knapsack for bandages, along with the two crosspieces that separated the poles and completed the stretcher in use. They could be told from afar, the stretcher bearers, wearing caps with bands that announced them. Because the regiment had plenty of bearers, the unharmed soldier must not help—he had been ordered not to help—a wounded comrade out of the line. Ambulances followed the brigade, one for about every hundred or two hundred men: a light, covered cart with springs, drawn usually by two mules, and carrying two wounded reclining, more when they could sit. Vehicles with medicine and other medical equipment and supplies followed in numbers according to those being served.

The brigade, too, had its chief surgeon. The commander, preparing for battle, would inform him of its likely course. The chief surgeon in turn ordered the siting of intermediate stations and field hospitals. Surgeons, those serving the troops, designated their forward dressing stations and reported the sites to the intermediate stations. Thus the medical service tracked and served every clash of arms. Organization had been assimilated; the system produced exactitude of procedure, and everything worked with few questions and little fuss.[2]

Let me not tire you with specifics. Allow an incident to illustrate the wake of a battle: May 2 at Chancellorsville.[3] I saw how they handled the wounded there. Jackson himself had been wounded. Stuart in his place arrayed the corps. The double line advanced, sharpshooters out front, a weak reserve in the rear. Destination: Fairview Hill, bristling with defenses. Mission: storm it in the face of that artillery and at the scowl of those earthworks, into the mouths of the cannon, into the teeth of blades and bullets. Stretcher bearers went with the troops of the first line, approaching the enemy to about five hundred feet. The bearers took cover in a ravine while the lines drew up for assault: one of the bloodiest. Quadruple rifle and artillery fire beat them back twice with casualty upon horrible carnage. Then on the third, the determined try, the Confederates took it with bayonets. The wounded lay in heaps that seemed to cover the hill. Even as the victors pursued the driven enemy, Death reaped His bloody harvest.

Hill taken, the wounded—those who could get up—got up somehow and went to the rear, wobbling, rifles butt-down as crutches, or crawling, or the head clamped in the crook of the arm. How often we saw it on those killing fields! Meanwhile in support of their units the stretcher bearers sprang into the thick of it, first to aid those who could not move. Using the stretchers I described above, or interlocked hands, or even a rifle, they brought their wards to the rear. Virginia's forested region abounds in creeks. Slowly, cheek by jowl, the throng staggered, shuffled, and groaned to the nearest. Nature herself marked the spot: Put the dressing station here, She said. The wounded took seats on their own if they could. Others the bearers placed so the victim's useable hand could reach water to quench thirst and bathe wounds. As almost every Confederate soldier knew how to apply a dressing, the moment the battle ended, the reserves did all they could to help with healing and nursing. In less than eight hours, except for the few stranded and alone in thickets, the wounded—every man hurt and in need—had been brought together for first aid. Perhaps twenty thousand lay there, not only ours but the enemy's, too, on the field. He had to leave his to our care.

Something water-resistant—a piece of tent, a blanket, even a shawl, on poles, trees, or guns—sheltered each wounded man from sun or rain. Every station therefore looked like a campsite stretching on and on along the creek and snaking into the distance. The place could also be compared to the colorful settlements of gypsies. Fighting-unit surgeons worked examining the wounded, dressing and bandaging wounds, and changing what had been applied crudely and in haste. Men named to move the severely wounded to intermediate stations would be moving them. These stations had to be on roads, being also ambulance stops. Here the chief and his brigade surgeons operated on easier cases. If hospitals could not be easily reached, they amputated too. Typically, however, broken bones got splints, and hard cases interim care, then on to field hospitals. These, usually near railroads, consisted of tents to shelter the worst until they could travel. The corps' most skilled surgeons did major work there.

Then patients went by rail to various parts of the South, according to plan. I mentioned the system at the start of this chapter. In a South of passionate patriotism the dispersal system proved popular everywhere. It also produced the happiest results. Statistics show that the wounded got well in astonishing numbers.

Aside from wounds, health prevailed in the Confederate army, thanks to the simple way of life. It may have aided the wounded's recovery rate. The Army of Northern Virginia forbade whiskey, as is well known. Southern soldiers drank only their tasty spring water, besides (instead of coffee or tea) a brew of roasted corn or wheat. Continual marching and other exercise kept the bloodstream clear and the blood from becoming too much for the vessels. No excess flesh on those men, either, merely what remained after incessant exertion and repeated exhaustion. Never did I see a Southern soldier overweight!

Despite opinion to the contrary, prisoners and wounded prisoners enjoyed good treatment, the one no less than the other. I saw it and proved it to my satisfaction again and again. Since I have failed to persuade some of my colleagues, I quote from the splendid account of Northern medicine. Assistant Surgeon Bacon, after complaining of little clothing and few provisions: "It may be said, in extenuation of this treatment of our wounded by the rebel authorities, that it was chiefly owing to their own want of food, clothing and other necessaries, and that we received from them much attention and courtesy." Assistant Surgeon Adolphus on the same day: "The major part of the wounded in the engagement of Friday fell into the hands of the enemy, and were well cared for. The wounded of the battle of Chancellorsville also remained in the hands of the enemy for the space of ten days, and were well treated, though their medical supplies were not abundant. . . . The rebel surgeons, however, operated as early as possible, and with good success so far as our men were concerned."[4]

The same can be said—"good treatment"—for Southern wounded: well fed, well looked after in the North.

### 3

Care had been organized on a grander scale there. They had the means to begin with, plus larger armies, more sickness, and greater numbers of wounded. In addition, for the first time, the army had at its disposal volunteers organized to help care for the sick and wounded. But, as on Southern, I shall not be detailed on Northern medicine. Allow me another example of field hospitals and battlefield medical service. I choose the last battle on an open field: the Wilderness, May 5, 6, 7, 1864. I quote from the clear and detailed report of Colonel and Surgeon Thomas A. McParlin, Medical Director, Army of the Potomac.[5] See also a map of this battle.

We join him after the North has reorganized the service around brigades and divisions. He has expressed satisfaction with the latest regulations, and delight at independence for the medical staffs of those units. Before hostilities resumed in 1864, he says, "all of the ambulances were thoroughly repaired, painted and marked with the distinctive badge of their several divisions and corps, details of medical officers and men for the ambulance service were made, and the persons so selected were carefully examined. As was to be expected, a large portion of those first detailed were rejected, regimental commanders having attempted to rid themselves of their weak and worthless men. The men attached to the ambulances were carefully and regularly drilled, minute inspections of everything connected with the ambulances and horses were made, and guidons and hospital flags were procured and distributed. The results of the labor and preparation thus expended will appear in this report. Tens of thousands of wounded men have been carefully, speedily and safely transferred from the field of battle to the field hospitals, and from thence to the large depot hospitals, and this has been done without confusion, without hindering the movements of the army or conflicting with the operations of the other staff departments. Closely connected with the ambulance system, and, to a great extent, dependent upon it, followed [sic] the organization of a system of field hospitals. The amount of transportation allowed for medical purposes was three wagons to each brigade of one thousand five hundred men, and one wagon for each additional thousand men. This allowance proved ample. The amount of regimental hospital property to be transported was reduced to a minimum, all the hospital tents and stores being turned over to the division hospitals. The medical staffs of these hospitals . . . [were for each] one surgeon in charge, one recorder, three operators, each with one or more assistants, and one medical officer to provide food and shelter. As the plan of the division hospitals varied somewhat, a brief sketch of the hospitals of the first division of each corps will perhaps best illustrate their peculiarities."

We will quote him on only one.

"In the 1st division, Second Corps, twenty-two hospital tents, fourteen army wagons and four medicine wagons were allowed for medical purposes, the division containing four brigades, twenty-one regiments and eight thousand men. Six of the army wagons carried the regimental property; four, the brigade supplies; two, the hospital tents; one, the cooking utensils and three hundred rations, and one was loaded with blankets, beef-stock, whiskey, chloroform, bandages, lint, etc., etc. In pitching the hospital, no attention was paid to brigade organizations, except that an operating table was established for each brigade, the corresponding medicine wagon being drawn up beside it, and the surgeons-in-chief of brigades were *ex officio* the operators. Thirty-six regular hospital attendants were employed in the preparation and distribution of food, dressing wounds and care of the patients. These men wore on the left arm a half chevron, composed of a green and yellow stripe. During a battle, or series of battles, the drum corps of the division, numbering three hundred and fifty men and boys, were put on duty in the hospital, being organized into five companies, commanded each by a sergeant, and the whole commanded by a lieutenant, having an orderly sergeant as an assistant. From this corps, details were made, whenever called for by the surgeon in charge, for pitching and striking tents, loading and unloading wounded, bringing water and wood, burying the dead and for police duty. A provost guard was present at the hospital during an engagement for the purpose of arresting malingerers, etc. One medical officer followed each regiment into action, the surgeons, with the exception of those in the division hospitals, remaining at the advance ambulance depot, which was usually about five hundred yards in the rear of the line of battle. When a large number of wounded were brought in, these last were sent to the division hospital to act as dressers. . . .

"In the Cavalry Corps, no system of division field hospitals was organized, owing to the peculiar nature of their service; their hospital train was, by order or by circumstance, seldom near enough to be available during an engagement, and was entirely absent during their long expeditions and raids. The usual operating staff was detailed, however, and a full supply was carried in their train; the nearest available house was used as a hospital, and the wounded were, in many instances, subsisted by foraging, as will be seen by the reports of the cavalry raids."

On May 4, McParlin continues, the cavalry on the Rapidan crossed on pontoon bridges. "The only means of transportation allowed to cross the river were one-half of the infantry ammunition trains, one-half of the ambulances, one medicine wagon and one hospital wagon to each brigade, and the light spring wagons and pack mules belonging to the various corps and division headquarters.... On the evening of the 4th, the Second Corps, with the reserve artillery, was near Chancellorsville; the Fifth Corps, near Old Wilderness tavern; and the Sixth Corps, on the heights south of Germanna Ford. The distance marched was about fifteen miles, the day was cool and pleasant, the men fresh and in good spirits, and there was but little straggling. Fifty rounds of ammunition, three days' full rations in the haversacks and three days' short rations in the knapsacks were carried by each soldier. Three days' rations of fresh beef on the hoof were also taken across the river.

"Early on the morning of the 5th of May, it was found that the enemy were advancing from Orange Court-house, with the design of striking our column at right angles while in line of march. The original intention of moving farther southward was temporarily abandoned, and the army moved into line of battle. The Fifth Corps held the centre, lying across the Orange Court-house pike, parallel to and one mile east of Old Wilderness tavern. Water for the hospitals was obtained from excellent springs in the vicinity. Tents were pitched, operating tables and kitchens prepared, surgeons and attendants were at their posts, and everything was in readiness for the reception of the wounded an hour before the cases began to arrive. The advance depot for the ambulances was near the turnpike, about four hundred yards behind the line of battle. A surgeon, with attendants, was stationed at Old Wilderness tavern for the purpose of dressing the slightly wounded who might be straggling along the road, having passed by the advance hospitals without being cared for. The wounded began to come in about twelve o'clock M. [meridian or noon], and by nine o'clock P.M., twelve hundred and thirty-five men had been received, fed, dressed and sheltered.

"The Sixth Corps was posted on the right of the Fifth, extending to the river, the second division moving during the day to the left of the

Fifth Corps. The hospitals of the first division of this corps were at the Spottswood house, on the Germanna Ford Turnpike; that of the second division, on the Old Wilderness Run, near Woodville Mine, and that of the third division, near Old Wilderness tavern. About one thousand wounded were brought in during the day, the greater part from the second division. The Second Corps got into position during the day along the Brock Road, leaving an interval of nearly two miles between its right and the left of the Fifth Corps. This space was for the most part filled up during the day by General Burnside's command and one division of the Sixth Corps. The hospitals of the Second Corps were located near Carpenter's house, one mile southeast of the junction of the Germanna Ford and Chancellorsville plank-roads. The site was a good one, with good water and two ambulance roads leading to the front, which was only a mile distant. About six hundred wounded were received during the day. The trains of the army crossed during the day at Ely's and Catharpin Mine Fords, and moved one part to Chancellorsville, the other to the vicinity of Woodville Mine. The first division cavalry crossed after the trains and moved to Oldrich's, picketing the roads towards Fredericksburg. The other divisions of the corps engaged with the enemy's cavalry, during the day, in the vicinity of Todd's tavern, where all the wounded were dressed and fed, and the necessary operations performed. They were then removed to the tent hospitals of the corps, one of which was established near Ely's Ford; the other, near the hospitals of the Fifth Corps. During the day, army headquarters were on a little knoll near Woodville Mine. During the 6th and 7th of May, the battle of the Wilderness continued, the corps and divisions remaining in nearly the same relative positions. As has been well said, 'this was a battle which no man saw or could see,' fought in the midst of dense thickets of second growth underbrush and evergreens, rendering the use of artillery almost impossible, and compelling the opposing lines to approach very near in order to see each other. It was a series of fierce attacks and repulses on either side, and the hostile lines swayed back and forth over a strip of ground two hundred yards to a mile in width, in which the severely wounded of both sides were scattered. This strip of woods was on fire in many places, and some of the wounded who were unable to escape were thus either suffocated or burned to death. The number who thus perished is unknown, but is supposed to have been about two hundred. The stretcher-bearers of the ambulance corps followed the line of battle closely, and displayed great gallantry in their efforts to bring off the wounded lying between the lines, but with very little success, it being almost impossible to find wounded men lying scattered through the dense thickets, and the enemy firing at every moving light or even at the slightest noise. The hospitals of the Second and Fifth Corps remained stationary throughout the battle; those of the Sixth Corps were collected on the night of May 6th, and moved to the vicinity of Dowdall's tavern, on the Fredericksburg turnpike. This removal was necessitated by a fierce attack of the enemy upon the right wing, which forced back and threw into temporary confusion the 3d division of the Sixth Corps. Some shells were thrown into the vicinity of the Spottswood house, injuring two ambulances, but doing no further damage." The army lost 15,004 men; and for every 16, one officer. On May 7 the evacuation of the wounded took place: to Washington."

Dr. McParlin also relates the oddity of "a somewhat interesting fact bearing upon the character of the conflict." The chief ordnance officer stated that "during the three days' fight," the army used "but eleven rounds of ammunition per man."[6]

McParlin's description lets you compare Union and Confederacy on transporting and caring for the sick and wounded. You see the resemblance. They differed in the later stages: during hospital care, convalescence, and recuperation. The Union—rich by contrast to the Confederacy, besides having many more cities—met the needs of treatment and nursing by concentration in general hospitals and in a manner opulent by Confederate standards. The Union thus keep patients together, hence their "general" hospitals. The Confederacy, an agrarian nation and short on everything but space, had to disperse them by quartering on the populace and the individual farmer.

Sick and wounded, concentrated in hospitals or other centers, harbored hazards and courted danger. Such plethoras could cause the worst of outbreaks, epidemics, among one's own populace. Patients must be given quarters as ventilated and salubrious as possible. Most of you—I take the fact for granted, colleagues—know of the North's airy, easy-to-construct patients' quarters. Though always attached to hospitals, they took different designs: either radiating from them or in geometric shapes also centered on service and administration. I list some of the better-known hospitals and the shapes and sizes of their patients' quarters.

- Sedgewick (circular) in Greenville, Louisiana, 1,500 patients;
- Lincoln (three-cornered) in Washington, 1,240;
- Mower (square), Philadelphia, 3,000;
- Hicks (half-circle), Baltimore, 1,100; and
- McDougall (horseshoe), Fort Schuyler, New Jersey, 1,660.

Partly public, partly private monies—taxes and contributions by organizations and individuals—built them. Models of modernity, they had everything the latest technology could provide. Only think: 300,000 dead by sickness and battle injuries in the Union forces. Assuming about 1 died per 8 sick and wounded, the North in four years fed, treated, and healed perhaps 2.5 million.

### 4

Such results demanded intensive, conscientious measures besides effort upon strain after struggle. America applied knowledge and know-how, the one encyclopedic and the other prodigious, to restore and heal those who sickened and bled. Caring for them in turn aided the well-being and prosperity of the nation. No wonder such an example must resonate in the armies of Europe. Thus we see in our military-medical care the outlines of the American way to handle and move the sick and wounded. Not only have we in Germany followed the American example in transportation, but also the Union and Confederacy in dispersal or quartering for convalescence as well as concentration in hospitals and recuperative centers, respectively, as applicable and possible.

True, our medical service does not approach the Americans' in size or excellence. Bear in mind that there can be no such achievement in peacetime. The medical corps then, and on the budget for peace, can be little more than needed in peacetime. Physicians and medical staffs then must suffer the fate of excess: fat to be trimmed. Yet medicine in the tranquility of peace must spare nothing to prepare for the travail and turmoil of war. The organism has life—let alone strength—only when stimulated. Idleness means atrophy; atrophy, death.

In order to get anything like America's excellence, we in peacetime had better organize medical detachments. This select and separate cadre would be schooled and trained, fully and rigorously, as soldiers as well as medics: to be the six or eight needed as stretcher bearers for each battalion. They would march with their units but unarmed. Fighting men thus lost will be more than replaced when commanders can insist with draconian severity that no soldier leave the line to help a comrade to the rear. As medical detachments at crucial and dangerous moments should spring to the aid of their units, so should the medics learn to feel as one by staying in touch with them. They must go along on maneuvers and be with their outfits in what they do in the field. Half a detachment per division could be mounted or in vehicles then.

The American physician enjoyed the distinction of major's rank. He deserved it. Troops elected other officers; therefore, some had little education, some learned their craft in the ranks, and neither group might be better than common soldiers in professional training and military science. Troops did not elect physicians. They had been to school, trained and tested there, and passed. They brought the army a wealth of useful knowledge that must be compensated in all fairness. They got their rank regardless what the troops thought.

Our medical corps distinguished itself in our last wars. Thus as in America, war made the corps effective. Its officers have been given ranks equal to those in other branches. (Some disparities prevailed for a time. They have disappeared.) Pay must be equivalent, too, that this young branch regenerate itself from the best after the ravages of war and the openings for recruit physicians be filled with the competent. Better paid, the medical officer will not seek civilian practice. He can serve the army and enjoy the prestige of his civilian counterpart. Thus a branch of the army be elevated and the army improved.

The corps wants to extend the terms of older physicians, and if possible expand itself to include an academy to train new ones. Who of my colleagues takes the wish to heart? You who know it could be *you* with the bullet-smashed arm or leg!

American generals acknowledged a truth you, colleagues, ought to live, swear, and die by, especially you who hold influential positions in wartime. I mean that the care of the wounded and the morale of the warrior go hand in hand. Greece and Rome exalted the wounded— Napoleon glorified the wounded—in the American South, women for the first time nursed and coddled the wounded—and the nation hailed the fallen as heroes. Who can wonder that out of such a state of mind came youth without fear to charge enemy lines and be indifferent to wounds? He who would evoke heroism from the armies must look to their welfare after battle. Every work of mercy, done then, plants a seed that thrives in future battles, often bringing forth laurels. Our lords and masters in the medical service—make no mistake: I mean the Americans in their Civil War—have set a shining example. May our commanders follow it, one and all!

# 8

# NAVIES

*The Americans confronted the issues foursquare.*

Introduction—Development of the Union Navy—Blockades and Blockade Running—An Example—<u>Monitor</u> versus <u>Merrimac</u>: Battle of the Ironclads—Wooden Ships versus Ironclads in Mobile Bay—Combined Operations—Capture of New Orleans—Raiders—Concluding Remarks: What Ought European Navies Learn from the American War?

The United States of America, a seafaring nation and world power superabundant of resources, fought a civil war that encouraged good but quickened evil to the utmost on a continental scale. Such a conflict must influence sea power elsewhere. Indeed, seafaring peoples had been converting from sail to steam and turning their thoughts to fighting with the new ships. Now and then they pondered the issues hit and miss, mostly miss. Artillery, for example, went down to the sea in ships more and more now—what about iron on hulls against those missiles then? Europeans had been paying attention to armies and army ordnance, so they could not be more alert when naval warfare grew critical. Meanwhile the Americans confronted the issues foursquare. The Europeans therefore watched the Americans become the first to float ironclads and make them fight. Below we shall see North and South confront the issues and together deserve credit for resolving them.

## 2

Only the North interests us, however, when we turn to our first topic: naval development in general. New Englanders—so-called after their region of towns along the northeast coast—elaborated ships, in peacetime followed shipping and pursued trade, and owned fleet

**117**

after commercial fleet. New Englanders especially, empowered by intelligence and industrial strength in their seaside towns, fomented and consummated the naval revolution during the Civil War. I mean that the maritime success of New England, together with the energy of naval leaders, expanded Union power afloat until, soon, on the rivers as at sea, the Stars and Bars could not challenge the Stars and Stripes.

A Union well armed at sea at the start of the war? People usually answer yes. In fact the Union took an unprepared, poorly armed, undersized navy, exerted on it a vigor of enterprise, committed to it the nation's inexhaustible resources, and revolutionized it as the war went on. Let statistics show why the result could only astound Europe. The Union had 42 active warships and 7,600 sailors in March of 1861. With 30 ships in foreign ports, how many could the secretary of the navy order into action at home? He could put 3 on line, plus 1 revenue cutter, and their 260 sailors. Unimpeachable numbers—proof positive of weakness. And then civilians and not naval officers conducted the first operations on the water! They tried with the president's approval to get men and matériel to Fort Sumter after Beauregard besieged it. The civilian steamer *Star of the West* tried to reach it. Charleston's batteries hammered at her. She had to turn back, mission failed.

It angered New England to be repulsed. The desire, natural in New England, seized the North: they would master the South at least at sea and whatever the cost. The government acted at once. In four months, keels for eight steam-powered sloops had been laid. Twelve larger steamers, big enough for two to nine guns, had been bought and another nine leased. Twenty-three gunboats, totaling five hundred tons, took to the water and struck the Confederacy wherever they could.

The rush to naval readiness doubled after Bull Run. The Union determined to hit and gain by sea what it had missed and lost on land. It bought another 121 ships, built 52; and of the mothball fleet, albeit half sailing-craft, activated 76. In the statistics of December 1862 we find 427 ships of the line, 363 more than in June 1861, totaling 1,577 guns. This buildup, begun with a bang, did not end with a whimper. Numbers at the end of the war: 671 of the line, including 71 ironclads and 37 large steamers armed with heavy Dahlgrens, plus 121 used only for transport. This astonishing energy and astounding output, this power upon power to produce, had never been seen before. Such numbers, getting what they merited, caught the attention and attracted the interest and won the admiration of Europe's maritime powers.

The navy's effectiveness grew with the numbers. But the navy did not have to defeat its likes on the seas. It must beat coast artillery, shallow waters, torpedoes, and the hazards of inland navigation.

### 3

As soon as numbers became respectable, the navy undertook its first mission: blockade. Much did the North put into it, aiming to cut off imports and cripple by sea the enemy dependent upon factories abroad. Northern ships also captured nearly all coastal fortifications and seaside military districts, ever narrowing and further constricting the Confederate range. Inland the North little by little took the vast Mississippi system, prelude to capturing the West and then conquering the Confederacy itself.

The blockade demonstrated the difficulty of blockades nowadays when steam empowers and emboldens the runner. The North threw about all it could into stopping him. Yet a runner mixing circumspection and sangfroid could penetrate with relative ease the porous ring of ships on station to block him. When we lay at anchor in the Cape Fear River outside Wilmington, North Carolina, in September of 1863, we saw the runners go in and out: seventeen steamships in, eight out, and not a shot from blockaders. Seventeen blockaders never suspected a thing. Built in Britain for the purpose and called blockade runners in English, they could go faster than steamers of the blockade. Painted greenish white, the runners could hardly be descried amid the foam and spray of the ocean and the sky's bright cast. They projected nothing like masts, only one of the shortest of cargo spars. Stacks, designed to dissipate smoke, for the same reason threw no sparks. The highest quality coal fired the boilers, hence a minimum of smoke and sparks to begin with.

Sand gathers outside about all North American harbors, in semicircular bars, at depths of five to twelve feet. Markers and buoys show the channels by day; ships triangulate on beacons by night. (A map of Charleston will illustrate what I mean.) During the war these aids disappeared, of course. The captain must find his way by soundings. He, the runner on his dangerous mission, must zigzag among blockaders, penetrate the blockade, find a channel, and follow it into port—all by the dark of the moon. Blind channels stop in the middle of a bar. Should he ground there or on a bar itself and so stick as to fail to struggle free by daybreak, nothing could save the ship. Many lifeboats dangled along the decks. The crew could but resort to them and save themselves and as much as possible of the valuable cargo.

It took a cold-blooded captain to win at this derring-do, especially when evading blockaders. True, the fastest blockader could not go as fast as the slowest runner. Yet he dare not speed up, even under fire when running the gantlet, lest he be in a blind channel. In a true channel his ship had nothing to fear; the pilot knew the way, be it daylight or dark. He raced into port then. In the dark his pursuers soon lost every trace of the fugitive.

### 4

Let me take you through the blockade as we did it into Charleston and out of Wilmington. Each time we had to evade seventeen steamships. The story begins in Nassau in the Bahamas. I had gone there after a while in New York.[1] Trade with the chief Confederate ports flourished in Nassau, through the blockade. The city's every corner hummed and bustled with it. I took the first runner out, the *Flora,* built in Glasgow, twin-screwed, steam-powered, and fast. She carried nothing but shirts and other clothing, precious cargo for Charleston. Two Union cruisers having been spotted, we went southeast among the Bahama reefs. Beautiful weather, dangerous route. After twelve hours and passing San Salvador, as it happened, we turned north to the latitude of Charleston. We kept our distance; the seventeen on the blockade did not see us. So far the tense moments had been at daybreak when dawn could conjure up an enemy that had closed on us since dusk. We stayed thereabouts but on the high seas until the dark of night. Carefully we went in but, reckoning false, missed the port. By soundings we had to feel our way south along the shore—and chanced on the blockade, nearly ramming the first ship. Had their men on watch fallen asleep? We came alongside. A shot crossed our bow. It flustered the captain. "Full speed ahead!" His order drove us with all our might among the bars. Of course in a few minutes and with panache we ran aground. Rockets and other signals had alarmed the flotilla, but, weather stormy, they could not spot us by their light. Nor could blockaders follow us into the shallows. They fired and missed, wasting their powder and squandering their effort. The tide favored us. After an hour—an anxious, dangerous wait—we floated; our powerful screws freed us; we could move. We spent the next four hours feeling the way among the bars, often about to bump into blockaders. They kept shooting. At last we got into a channel. The pilot recognized it after sounding a bit and taking us about a half hour along it. Yet we did not proceed and clear it. The captain had been urging the ship along; now he stopped, reasons unknown. About to start, persuaded then with difficulty to wait until morning, he held the ship there. Meanwhile if necessary we would repair to our many lifeboats and land in them. Early in the morning we learned— with such joy!—that we had anchored facing the bar, the chief of the sandy reefs that block this and almost every port here. The captain had but to turn and run in. Fort Moultrie gave our Confederate flag the friendliest of salutes.

Leaving the Confederacy, I shipped out of Wilmington on the fast and famous steamer *A. D. Vance,* a Confederate navy captain commanding. Overloaded and off balance, she dug with her keel her own difficult channel through the bar, running aground again and again. At last she cleared. The pilot took course and corrections from shoreside lights while people ashore read the course from the light at her stern. Thus they tracked us. Should we ground, they could signal the way out. After a stormy passage of three days, we put happily into Saint George in the Bahamas, undetected.

Reports out of the North give the impression of an effective blockade. Recall first how Americans exaggerate, a nasty habit. They exaggerate and exaggerate. Official reports of the blockade exaggerate. The blockaders could have had no idea of the many runners in and out. They ran by night. At dawn they had long since disappeared beyond the horizon or into port. In 1863 I saw the fleets of runners in Nassau and Saint George and in Charleston and Wilmington. Anyone who saw them knows with me that despite the commitment and exertion of Northern naval power, the blockade succeeded only against ships of sail. They dared not try except under the most favorable circumstances. Official records state that the blockaders seized 153 vessels up to the end of 1861. After detailed, meticulous inquiry in neutral and Southern ports, everything I learned says many of them must have been small coasters of one kind and another. True, later—I know, being in the thick of things in 1863—the much-tightened blockade had become more successful. Yet the capture of one runner constituted an event to be discussed in every Southern port and at length. Captures amounted maybe to one every month or two, all told.

Having made those claims, should we be tempted that the blockade did not harm the South? The issue remains clouded. Blockade running spelled d-a-n-g-e-r. Thus the blockade interdicted shipping not by the fact but by the threat of capture. The threat multiplied the cost to get in or out. Those expenses bloated to an enormity that in time sank the South. Yet the North did not begin to seal the South in fact until the North took the shoreside fortifications that commanded the mouths of harbors. From that moment, of course, entry into any such harbor proved tough. Still, when Sherman's campaign forced the evacuation of Charleston, a steamer had arrived only the night before. It had run up the Ashley River after passing hard by the hostile Battery Gregg.

I have discussed the blockade in detail because we in Germany think we could be easily blockaded and not able to force our way out. So I emphasize: measures may be extreme and not produce a blockade that seals the coast. Enterprising sailors will be found everywhere. They, in hope of lucre, will take risks and run blockades.

5

The blockade amounted to a passive assignment for the fleet. On the high seas, along the coasts (other than blockading), and on the rivers, the fleet played an active, aggressive part. In other words, in carrying the war to the enemy, it did not merely figure in; it *mattered.* In 1862, for example, it had to provide the transportation and firepower that sustained the peninsula campaign under McClellan. Moreover, that campaign saw the inaugural battle between ironclads; and that battle made the campaign historic. The apparent became obvious; a likelihood became a certainty: in naval vessels, wood must yield to iron.

Before being called to move McClellan, the fleet lay in Hampton Roads, keeping watch on the fortress at Norfolk: three of the five big warships off Newport News, the other two perhaps near Rip Raps.[2] It had been heard that a Lieutenant John M. Brooke as early as June 1861 had armored the hull of the *Merrimac* and added armor to superstructure as well. I mean that the Union fleet had heard but not believed they might meet an enemy their equal. The ships therefore lay anchored and unprepared. Signaled early on March 8 that she had left Norfolk, they worked with the aid of tugs to weigh anchor and get into action. The *Merrimac*'s armament consisted of a ram and ten pieces of light ordnance.[3] She moved at a slow five knots and looked like a house submerged to the chimney and roof. She and her escort of six light gunboats went straight to the struggling Union vessels. They fired at her but missed. Nearing, she gave one, the fifty-gun *Congress,* a load of grape and then every gun of a broadside. When two returned all they had, the missiles bounced off like rubber balls. Headlong, headfirst, she drove at the *Cumberland* (twenty-four guns) still at anchor. The ram hit the target at full power, punched a hole in the starboard side, and drove her to the end of her chain. Fires broke out aboard, she began to sink, yet the brave crew shot back. Not to be saved, the *Cumberland* went down but for the top pennant. It marked a few feet above water the spot where she struggled. The crew saved their skins by swimming and in boats.

Having knocked out that opponent, the *Merrimac* lunged at the *Congress.* This Union vessel had to fight alone; her allies could not approach in the shallows. The *Merrimac*'s escort (on the other hand) supported the *Merrimac* at long range when she opened at 250 paces. The *Congress* caught fire; the crew must abandon. About midnight she blew up. The second opponent had been knocked out.

The *Merrimac*'s captain, Franklin Buchanan went for a third, the *Minnesota,* aground. Unable to get near enough to ram in the shallows, he had to be satisfied to fire until nightfall. The winning ironclad suffered a mere two dead and eight wounded besides minor damages that did not prevent fighting. But the ram had been bent so as not to be worth using until fixed.

Had fate prepared a sad end for the Union fleet the night of March 9? If running offered the sole hope, what of the *Minnesota,* still stuck and helpless? At this baleful moment, unexpected help: the *Monitor,* readied during the fighting, an ironclad with a turret and two eleven-inch Dahlgrens in the turret. At 2:00 A.M. she steamed in. Near the stranded *Minnesota* she took post where the *Minnesota* would hide her from Norfolk. The *Merrimac,* to Norfolk for repair and supply, returned at 6:00 A.M., suspecting nothing. She went for the *Minnesota* and learned what the *Minnesota* had hid from her: an enemy her equal.

The armored monsters fought their astonishing, electrifying, hair-raising sockdologer of battle. Point-blank they fired. Each, whenever gaining the flank, tried to ram, to drive iron into the other's guts. On one of the passes the *Merrimac*'s ram broke off. The battle renewed in fury, neither able to gain much. A shot to the pilothouse nearly blinded the *Monitor*'s commander, Lieutenant Lorimer Warden. The *Monitor* withdrew. The *Merrimac,* also hurt, gave up at the same time and went back to Norfolk. Not even the stranded *Minnesota* tempted her now. Spectators crowding and festooning the shores had seen a spectacle.

## 6

Iron had outdone wood. Iron would prove all the more superior ship against ship. That battle certified the superiority of ironclads and of course prompted more, beyond number, north and south. Little innovation went into them; successors aped the originals; the North modeled on the *Monitor,* the South on the *Merrimac.* Northern turret ships one-on-one usually enjoyed better-built machinery and higher-quality armor on hulls and around guns. European navies, minding that success, adopted the turret but modified it. I remain nonetheless convinced of benefits to armoring standard hulls and making every missile not only strike iron but also at an angle. Such rebuilding and refitting costs little compared to complicated turrets. As long as the navy must have iron, those benefits taken together vouch for iron-on-wood, at least for coasters.

True, ironclads seemed to hold the terrifying edge in ship-to-ship combat—for a while. But their power diminished, their dominance declined, when cogitation and experience produced guns and projectiles that could pierce armor. Wood also learned to outmaneuver iron and so well that iron could not strike back.

The famous Union admiral David G. Farragut proved what I have just asserted. Runners had been going free and easy in and out of Mobile Bay until the spring of 1864. He led the interesting assault on that stronghold the same summer. He must reckon with the stout Fort Morgan overlooking the harbor—after he dealt with the mines investing the harbor's mouth. Yet the armored ram *Tennessee* deserved more fear: a hull forty feet by two hundred, sheeted with three inches of oak plus another sixteen of fir and three of iron in vertical plates under two in diagonal and another in vertical. Farragut knew the courage and skill of her commander, Admiral Franklin Buchanan. Spies described this ironclad: not only strong and well armed (six guns) but also exceeding every predecessor in size and armor (six inches). Unfortunately, such bulk wanted power, but, as always in the South, the engines did not deliver.

Ought Farragut have been afraid? Should he at least have intimated fright? He said wood could attack the iron ram without fear. Approach within one or two hundred paces, he said. Pour in the broadsides. Send her to the bottom or drive her aground. Now with the *Tennessee* he did fear shallows that might prevent pursuit. (The fear would prove false. She drew twenty feet and must avoid shallows.) Therefore he brought four monitors with him. He had wasted no time; four enemy ironclads neared completion in the Norfolk yard.

On August 5 his flotilla, including monitors and steam-powered sloops, entered the bay, there to wreak havoc with guns. Not only asserting chains as armor against cannonading, they also had been fitted with iron rams. In addition a gunboat accompanied them portside—Fort Morgan lay starboard—to push them out of harm's way should machinery be damaged. The monitors arrayed themselves between the rest of the squadron and the fort and opened fire. "Fire" meant the gunners did all they could to deliver everything they could. So for a half hour they steamed past the fort at deliberate speed, making the inward passage with little damage except for the monitor at point. She hit a torpedo and took all hands to the bottom.[4]

In the bay the wooden ships separated from the monitors. On the Confederate side only the *Tennessee* and three woods could fight. The Union flotilla sighted her. The flotilla's woods closed in. The battle developed and with it tactics never used by any navy before. The *Tennessee*—underpowered, crowded by the attackers, and unable to maneuver well anyway—could not maneuver at all now. The big Union steamers began ramming with all their might. Steering badly, laboring to move, ringed tight, rammed time and again, she also absorbed broadside after vicious broadside. Yet she suffered no damage but wounds among the crew. The *Hartford* gave her everything of nine-inch Dahlgrens at ten feet. Even that barrage did no harm. Her guns could fire; she fired for an hour. The monitors came in with their fifteen-inch Columbiads. They shot away her rudder, stack, and part of a casemate. This pounding split her gunport covers, no less, at last. The brave ironclad's forcible and tenacious resistance stopped. Buchanan himself had been wounded; his ship must surrender. In turn, Mobile itself must end resistance.

## 7

Those episodes, *Monitor* versus *Merrimac* and the Battle of Mobile Bay, show the Union fleet dueling with Confederate ironclads, not a usual operation for the Union fleet. Almost always it must fight land forces and usually in cooperation with land forces. By those army/navy combined operations, the North captured the Mississippi, the key to the West.[5] In the end, because it cooperated with the navy, the Union army prevailed everywhere.

When the operations must engage forts or pass fortified sites, ships usually put on proper dress and fitted themselves for the part: the right-colored paint as well as engine rooms armored with chains or padded with bales of cotton stacked outside and soaked. Energy, daring, and luck must do the rest. Not to worry, the Union navy had all three and with them did many a heroic deed that deserves a place in the history books.

## 8

The capture of New Orleans shows how the Union conducted combined operations. We also see there the difficulties in such operations. The example recommends itself twice more: (1) The capture of the famous entrepôt exerted an influence on the war; (2) Farragut commanded with courage and discretion. Rarely will you find a discretion so extraordinary and a courage so singular.

Under him a fleet formed on the Mississippi below New Orleans. The city's downriver defenses, two small forts at a bend about fourteen miles from the city, stood against the approach. His thirty armed steamships—frigates, monitors, gunboats—and twenty-one mortar boats would attack and reduce them. The fleet must also deal with obstructions laid on and in the river. A boom, floated between the forts, could not (though chained fast) withstand objects swept along on the currents besides the currents themselves. Old schooners replaced it, sunk, chained together, and anchored as a block straight across. The obstructions must be removed and the defenses smashed. Then landing forces under General Benjamin F. Butler would attack New Orleans itself.

The first try, on the Lake Pontchartrain route, had been given up and the second decided on: straight in by breaking through on the river. With difficulty, Farragut and Butler brought the fleet over the bar and through Southwest Pass. They landed troops by way of Sable Island to address one of the forts, St. Philip. Two schooners, positioned in the Gulf to the west, could with their guns reach the other, Jackson. The attackers harassed St. Philip, approaching from the unexpected side.

The river had ruined the boom, the first obstruction. A storm practically wrecked the second. And then fireships had been readied—wooden vessels to be torched and sent as incendiaries against the attackers—nothing else remained to block the attack. They could not be steered—who would stay in the flames to steer?—they had to be sent down adrift. So, before the battle, meandering, not only did they bring confusion to the task, they also drifted against Fort Jackson's wooden bulwarks. Later, during battle, the bulwarks caught fire therefore and helped the attackers by giving a point to aim at.

To defend the river, besides installing obstructions, the Confederate navy next had seven common steamboats bulletproofed with bales of cotton and armored with chains. The big ram *Louisiana* would serve only as a floating battery from where she lay under construction. The monitors *Manassas* and *Morgan* could merely approximate the need; their engines lacked power to ram. A second line of defense, thrown up behind the forts, feeble digs, scarcely deserved to be called earthworks.

The Union fleet, under Farragut's circumspect leadership, made every possible preparation for battle: engine rooms armored with chains, and paint applied to enlist color on the Union side. Hulls dark, to blend with the shore, would create a poor target; decks white, for contrast, would let anything be easily found at night. Each captain also took his own measures for resistance and defense; they neglected nothing.

Southwest Pass verged on shallow. Luck favored them, however, and they cleared it and swept upriver. Two mortar boats could begin the bombardment on April 16. Gradually the rest joined in. By April 18 all mortars and a few gunboats could take part. Forests on either side provided cover to most; the other few had to anchor exposed in open water. Bombardment continued without pause until April 23. Confederate artillery, lacking the range, could return but little. Union gunners took Fort Jackson as the chief target. Barracks and wooden fortifications caught fire. The defenders put fires out again and again. Then the men must suffer the chill of the night, having soaked their blankets and coats to smother the flames. They suffered more when the river rose to flood the fort and they had to stand in water sometimes knee-deep.

As early as the night of April 20 the Federals tried to remove what remained of the barrier and open the river wide. They used a gunboat under cover of night and much ordnance. Luck favored them again. In the afternoon of April 23, Captain Theodorus Bailey could move the first squadron into action, six steamers and five mortar boats. They would also be the van against Fort St. Philip. The leader, the *Cayuga*, Bailey aboard, met the Rebels' riverboats opposite the fort. She did not hesitate; she attacked. After a short duel, she forced three to strike their sails. The *Varuna* and the *Oneida* arrived to support her. The Rebel monitors *Morgan* and *Manassas* rammed and shelled the *Varuna* to the point of sinking. She must steam for land to put the crew ashore. Her guns fired back nonetheless bravely to the end. Shots at close range hit her so hard she had to strike her colors. The *Manassas*, after disabling and almost sinking the *Varuna*, headed for the Union steamer *Mississippi*, intending to ram at full power. The *Mississippi*, turned aside, swinging at the apt moment outward in a curve into the channel. The *Manassas*, ironclad but already shelled until crippled, rammed the shore under a full head of steam and had to be abandoned. Refloated, she coasted downstream to continue to spread fear until her true condition came to light when she caught fire. Later she blew up. Other ships burning and sliding downstream set fire to one of Farragut's second group, the *Hartford*. With a struggle the crew saved her.

The Union fleet had fought its way upstream. At dawn, seeing both forts passed, Union sailors decided to proceed. With a few shots they crippled the Confederates' second line. Nothing stood between the Union and New Orleans, leaving New Orleans no choice. The Union could steam into an open city, but, as a precaution, two gunboats took the point position. They shelled both shores, drove off some of the pickets on guard, and captured others. On April 25 the fleet entered the port. Two days later the personnel of the two forts, those who could not escape, likewise gave themselves up.

## 9

In revenge for maritime losses, the Confederacy commissioned cruisers that would be independent raiders, for instance, the *Florida*, the *Alabama,* and the *Sumter*—the last being the best-known—and Raphael Semmes the best-known of the captains commanding them with daring and skill. The Union put fast steamers on the water—the screw-sloop *Kearsarge,* for example, and others like her—but had nothing to outrun freebooters built in Britain for speeds practically unmatched at the time. The Union could not catch them, did not lay hold until the Confederacy surrendered, so fast they ran. This shortcoming in the Union navy meant damage to Union shipping and losses to Union commerce. Privateers took 193 prizes, worth 13.5 million dollars. Harm did not stop there. It sowed dread that grew and flourished until no firm dared trust goods to the commercial fleet. It hurt so much that—in addition to the nation's warships and the gunboat fleets and the armed patrols—the maritime states wanted their own corvettes packing two or three strong guns and driven by engines so powerful they could overtake any merchantman like child's play.

The Union navy prevailed against the raiders. What it did, especially in the later years, being all of a piece, helped win the war. The raiders' successes, piecemeal, whether we score them one by one or together, meant little by comparison. Add, multiply, square, cube, what you will to enlarge them, they do not amount to much.

## 10

"In about a year the Union made itself a first-class sea power. Therefore in peacetime the world's sea powers need not maintain expensive navies." How many Americans did I hear say it? How many Europeans have I heard quote it? Perhaps, superficially understood, the Union navy's development justifies such a view that the navy can be created after war starts. But as experience has proved it false for armies, so a closer look at America's example proves it false for navies. Recall first that at the start of the war the Union had little to fear at sea from an enemy unprepared there. The Union could not be taken by surprise; the Confederacy had no fleet to do the trick. By contrast an unprepared European navy faces dangers unknown to the Union. True, furthermore, the Union's astounding effort did create a fleet. A glance at the record suggests an armada invincible. More than a glance says otherwise. The record consists of the Union's reports. The Union laid it on thick. Remember: Americans exaggerate. Trust them, have faith in the record, and you shall believe that at the end of the war the Union's navy could have swamped every fleet in Europe and not have thought twice of such an accomplishment.

Only appreciate the navy's state. It may look superior and the superiority seem colossal. Truly understood, the "superiority" withers. Except the 37 large steamers, few of the 671 ships could have fought overseas. The 71 gunboats could operate on the rivers and along the coast; the 112 sailing vessels would do as transports; but the worth of all ended there. The gunboats, even the frigates, would have met their match in British men-of-war. The rest, mostly smaller craft and converted merchantmen, raise doubts of value at best. They could be counted on as blockaders against their own kind unarmed, but nothing more. In short the navy lacked substance. Only the admirals can be called world-class for daring and resourcefulness developed in years of fighting.

Questions about the nature and quality of the Union navy suggest others about ourselves. Have European sea powers set a biased course in shipbuilding? Should we prefer superior guns to superior armor?

If we study battles, the *Merrimac*'s with the Union's wooden ships comes to mind first. They fell prey to her, no contest, but we must seek reasons beyond wood. In the first place the woods lacked speed to dodge the ram of the iron. Furthermore, unable to maneuver in shallow water, some ran aground and stayed stuck. Moreover, small guns even in ships of the line could not deal with ironclads. Finally, the management of the fleet lacked coordination. While half lay inactive near Rip Raps, the other fell victim to the ram of the *Merrimac*. Had the wooden fleet possessed what I say they lacked, they could have glared at the newfangled ironclad and said, "Come on, we're not scared of you!"

Farragut and Porter ran powerful land batteries with ironclads. They also did it with wood, armored of course and well armored at the points needing more protection in cotton bales and anchor chains to shield machinery against hits by big guns ashore. Farragut even said he dared and would not hesitate to take up the battle of wood against iron. Had he glimpsed the future? During his war and after, naval artillery and armor underwent many and costly tests. Artillery has proved superior; no armor can stand up to the best guns. Indeed, a medium-sized ship cannot carry the thickest armor and the heaviest guns. One must be reduced in favor of the other.⁶

How we should benefit, we Europeans, if we sloughed armor and betook ourselves to battle behind the offensive might of superior artillery on the best of ships of wood! What we should have learned from the American Civil War!

PART TWO

# THE SINEWS APPLIED:
# A SOLDIER'S HISTORY

# 9

# WAR ALL THE WAY: MOSTLY EAST

*The people of the North demanded the enemy's seat of power, Richmond.*

From the Beginning in 1861 to the Spring of 1862—McClellan's Peninsular Campaign, 1862—Pope's Try against Richmond, 1862—Burnside's, 1862—Hooker's, 1863—Invasion of Pennsylvania, 1863—Siege of Charleston, 1863—Grant's Campaign against Richmond, 1864—Sherman's March through Georgia, 1864—War's End, 1865.

In the first months of 1861 the northern and southern parts of this union pursued antagonistic interests. Discord caused dislike, and dislike became anger, felt by both sides, concealed by neither. People young and old and of every class on either side would give all and lay down life itself for their causes. Rebellion made good its threat. Civil War came to shatter the young aggregate of North American states. We in these military studies neither ought nor shall ask about causes, be they constitutional, political, or economic. Let it suffice for us that the war occurred because of how those free and independent states developed. A glance at the hatred on both sides, and our appreciation of the bitterness of the struggle, will do for what matters to us.

The party of the northern states, the Republicans, won the presidency for 1861–1865. The election released the animosity and the rancor of grudges held for years. Abraham Lincoln took office. Seven states left the Union. Secession by South Carolina, Mississippi, Florida, Georgia, Alabama, Louisiana, and Texas did not remain political long. War broke out. The Union's Fort Sumter guarded the harbor at Charleston. South Carolina attacked. The fort gave up on April 14, 1861. Itself insignificant, the incident struck sparks to ignite tinder

**135**

abundant on the continent. Lukewarm feelings caught fire. The undecided had to decide. When Lincoln called for volunteers, states must choose whether to go to war against sisters in rebellion. Virginia, Tennessee, Arkansas, North Carolina, Missouri, and Kentucky chose rebellion.[1] They and their confederates created the Confederate States and elected Jefferson Davis to preside from Richmond, the new capital of the new nation.

The enemies faced and studied each other a while, surprised at the turn of events and the rift. It took time besides to organize nations for war and forces to fight. Skirmishes at points of friction sufficed at first. The new belligerents irritated one another, for example, at the North's many forts along Southern coasts. They amounted to Confederate pushovers; only three small ones stayed Union. Forced also to evacuate Norfolk and Harpers Ferry, the Union burned those arsenals in vengeance.[2] The Union tried its first blockade: little success.

The skirmishes involved fewer than one or two thousand troops. Neither side tried to engage larger units until June of 1861. Both sides made gains and suffered losses. Southern cavalry, under General Turner Ashby's snappy leadership, here and there gave notice: We exist. Clamoring for larger enterprises, the people of the North demanded the enemy's seat of power, Richmond. Volunteers for the purpose enlisted for three months and joined 15,000 regulars, bringing the Union army to 70,000.[3] General Irvin McDowell led 53,000 armed with 49 cannon south to subdue the wayward. At a creek called Bull Run he met General P. G. T. Beauregard, 28,000 Confederates, and 49 cannon in a strong position on the right bank. The world knows how the battle ended. The South's General Jackson distinguished himself, winning the name Stonewall. The Northerners fled in panic back to Washington. The Southerners, also shaken, dared not a step in pursuit. What an outcome to an expedition begun in vainglory and bombast—how droll! It nearly extinguished the European military establishment's interest in the war that followed.

Green manpower could not produce decisive results. Heeding that precept as Bull Run's lesson, the sides paused with armies in Virginia and spent almost a year organizing, disciplining, seasoning, training.[4] Only in the West did the fires burn on. There the Union not only dominated in troops, speed, and will, but also turned the rivers, the Mississippi and its tributaries, to its naval advantage.

The Mississippi had cast over the region a net superb for floating batteries. The Union, arming a tiny riverine fleet with excellent cannon, weighed naval forces in the scale of battle. The Confederacy, with little waterborne armament in counterpoise, must equip forts with sometimes wretched artillery instead. Could there be doubt who would win in the West?

The northern line of this theater extended from West Virginia's Kanawha River and across Kentucky to Missouri: about 675 miles with only 60,000 Confederates to defend it. Worse, Kentucky, Tennessee, and Missouri retained military autonomy, controlled their militias, and did not take orders from the Confederacy—not at all.[5] They fought on their own, independent of either side, and without reference to one another. Those states for that reason suffered the fate of self-willed partisans. Intending to spare themselves, they became the anvil to both sides' hammers.

Confederates under General Lee in West Virginia met General McClellan.[6] Others, in the center as well as along the Mississippi, took on General Grant. McClellan forced them back to the Alleghenies, Grant to the Tennessee as far as Memphis by *coups de main* at Forts Henry and Donelson and *tours de force* at the strongholds of New Madrid and Island Number 10. In those bold, skillful actions the North joined land forces with gunboats in combined operations. Out west, then, defeats made the Confederacy draw in its flank to the heights of Memphis. Unsuccessful engagements one after another shortened the strategic wing to a more defensible spread.

We ought now note that those battles in the West, being ahistorical or anamolous in form, differed from those in the East and departed from our experience. Only in Virginia did the fighting resemble Europe's. Out west, for example, Grant and A. S. Johnston squared off near Corinth. The Confederates stormed Union positions the first day, to lose them the second to fresh, newly arrived Union forces. The two-day bloodbath remained undecided. Thus they did not fight in the West as they did in the East or we in Europe. Even the West's first big battle—Pittsburg Landing, called Shiloh by the Confederates—confirms the distinction.

Meanwhile, Confederate losses in the West paled to moral victories for the Union, compared to the Confederacy's disasters along its coasts. Again by combined operations, bringing in ships and first-rate artillery, the Union took Fort Hatteras and Roanoke Island with the upper hand: victories never in doubt. They provided the keys to Pamlico and Albermarle Sounds. The Confederacy also lost Fort Pulaski (guardian to Savannah harbor) and Fort Macon, North Carolina. Then came Admiral David G. Farragut. The audacity of that man, running the forts that guarded the Mississippi, and steaming full-ahead into New Orleans no less! The important loss of New Orleans brought dismay to the South. Worse, coastal successes empowered the North to tighten its blockade until only runners, none other than steamships bought for the purpose, could penetrate. The tighter the seal, the fewer the sources of clothing, weapons, munitions, even salt. The harder, then, for a South without factories.

The North wanted to begin at the coasts and little by little draw a noose tight and choke the South. But the terrain favored the South, in particular the nearly impassable swamps on the east coast. Elsewhere, if terrain permitted attack, it contradicted it by offering multiple places. Too many places to attack, too many attacks, diluted strength. The South meanwhile stayed strong with concentrated thrusts and got stronger with deadly ripostes. By the spring of 1862 the North had not begun; the noose hung loose. The South stayed Confederate. Secession sang "Dixie." The Union whistled it.

## 2

I have said that in the East, the main theater, a pause ensued after Bull Run. Each side, using the interlude, strove to turn masses into armies. Only with a trained force could McClellan implement his plan to move from the James-York peninsula and against Richmond. He for the Union, a consortium of generals for the Confederacy (former Federal regulars), turned their backs on one another, confronted not the enemy but their problems, and labored at solutions. McClellan finished in April 1862.

The famous duel of the first battleships, *Monitor* versus *Merrimac*, had secured the way for the fleet. He landed his invaders at Fort Monroe, about 110,000, and went toward the Confederate capital. General John B. Magruder and 11,000 opposed in light fortifications along the Williamsburg lines.[7] All Magruder's positions lacked men but some gained strength behind swamps. Magruder fooled McClellan into attacking them. The ruse gained time to concentrate forces and stiffen the capital's defenses. The Confederates brought troops and guns from every coastal point, abandoning even the fortified port of Norfolk, to unify and reinforce themselves to parry the advance on Richmond, the capital embattled.

At last on May 5 by choice the Confederates abandoned the line opposite McClellan. With prudent vigilance he advanced again toward Richmond. The fleet accompanied him up the York. He halted in the swampy valley of the Chickahominy, where the Confederates attacked. The battle of May 31 and June 1 at Fair Oaks (Seven Pines to the Confederates) proved indecisive, but he felt that other powers than the Confederates had crippled him. His army had dwindled to 90,000, for example, and a hostile government in Washington waffled in support.[8] Therefore he took every precaution. A supply base at White House on the York, protected by the fleet, would provide the security of warehouses and magazines. Did he need proof of vulnerability? On June 14 and 15, J. E. B. Stuart hit his rear. The cavalry assault—Americans said *raid*—destroyed the base. Did he want a clearer, sharper signal, proof positive? Jackson came out of the Shenandoah by surprise and supported Lee in the offensive. McClellan got the proof: defeat at Mechanicsville the twenty-sixth of June, at First Cold Harbor (Gaines's Mill) the twenty-seventh, and at White Oak Swamp (Frayser's Farm) the twenty-ninth.[9] His Union army could not exist on the York.

Damaged and in disarray, his troops retreated. Fighting continued. Arrived at last at Malvern Hill after that march of horrors from White Oak Swamp, he rallied with big guns of Union boats to repel the Confederates. Later he reached Harrison's Landing without hindrance or further harm, then by boat to Alexandria on the outskirts of Washington. His army had suffered catastrophe. He fell out of favor.

Jackson's movements, inseparable from McClellan's peninsular defeat, exerted decisive force that established Jackson's reputation. To appreciate him and what he did, we must go back a little. Lee had squared off before Richmond while Jackson angled out of the Shenandoah. By offensives head-on and from the flank they would shield the capital from the northbound storm. Jackson's division of 15,000 opposed a semicircle of generals and their units: a corps under Nathaniel P. Banks; another under John C. Frémont; the division under Franz Sigel; and, far to the east, Irvin McDowell's corps. Jackson's latest dash had brought Lincoln's order to keep McDowell near Washington for its defense. Thus McDowell had not reinforced McClellan against Richmond but stayed near Fredericksburg—useless to McClellan and the least danger to Jackson. At any rate, 84,000 enclosed Jackson on three sides and pressed him. He, in a master stroke, struck hard four times like lightning, then hurried south to the Chickahominy without losing a man, while the enemy backed in fright. Jackson arrived at Cold Harbor, as I have said: there at the moment to slam McClellan's right and seal the expedition's fate.

Meanwhile west, Grant and Porter and Farragut, forceful leaders, fought and fought and fought. They captured Memphis on the Mississippi and besieged Vicksburg. The army owed success to a navy as well as to heavy-artillery support in addition to engineering superior to their Confederate counterparts.

3

In the North the people wanted to fight in other ways and by drastic means. To the challenge a new star had risen, a luminary likelier to coincide with the fresh and grandiose notions of public opinion. General John Pope shoved from the north straight at the Confederate capital. The shove brutalized Confederates along the route, leaving them angry, confused, and desperate.

Again the peripatetic Jackson took on the invader. Feinting, skirmishing, engaging, disengaging, Jackson would hold long enough for Lee and the main army to move in. At Cedar Run, Jackson put up a fight, and Pope met the first resistance in the most interesting of their bouts. Jackson attacked larger forces, outcome indecisive. But the Northerners camped too near the battlefield, outcome decisive. On August 8/9, the night following the standoff, Jackson hit and routed them. Stuart and the cavalry joined in. Southerners brutalized by Pope had told Stuart what he needed to know. He circled Pope's army at night, nobody on Pope's side the wiser, hit Pope's headquarters, and scattered Pope's troops at Catlett Station. He missed Pope himself, who happened to be absent. Stuart's horsemen seized the papers and other things of headquarters and burned the stores intended for Union troops.

Stuart had gone ahead. Since chances could be taken against Pope the incompetent foe, Jackson followed Stuart hush-hush. Jackson, like Stuart, appeared where unexpected, at Pope's rear. He destroyed Pope's central depot after winning at Manassas Junction. Pope, greater in numbers, turned them around to get rid of the numerical inferiors harassing him. Too late. Lee had tracked Jackson and waited poised to reinforce him just north of Pope. This Battle of Manassas, or Second Bull Run raged for two days, enormous, violent. Pope tried and tried. The battle ended on August 30, 1862, when exhausted, he turned east and brought the wreckage of an army into the safety of fortifications outside Washington. The cautious McClellan had erected them. The road north lay open to the Confederates. They took it at once.

The reins of Union generalship hung slack. Pope had reduced the army to confusion; it clamored for McClellan. With patriotic zeal he seized the reins in strong, capable hands before the order from a perplexed government. Again he faced Jackson heading the Southern advance. Jackson crossed the Potomac on September 9, then moved on Harpers Ferry, a fortress of weapons, ammunition, and food. McClellan would have aided that hub with an army from all over the Union and reinforced by Pope's shambles. Could the hotchpotch, outnumbering Jackson, have defeated him? McClellan tarried too long. Jackson took the Ferry (September 15) and waited on call. McClellan in character had prepared with prudence, thereby stalling the Union drive as much as did Lee's clever countermoves. Lee and McClellan met on the seventeenth at Antietam (Sharpsburg). Jackson had reinforced Lee.

Whenever the Confederates left their soil they lost their nerve. Out of the Confederacy, out of courage, as if a phantom swiped it. Now, in Maryland, they fought on the defensive, courage gone to the phantom. Though excellently led, they did no better than a draw. Lee felt too weak for the offensive. The nineteenth saw him already back across the Potomac. The rear guard, in the skirmish at Shepherdstown, West Virginia, on the twentieth repelled an attempt to pursue them. A pause; the campaign had been exhausting. Stuart used the moment to go north again with two thousand picked troopers the length of Maryland, the swift cavaliers destroying supplies and confusing intra-army movement.[10] They would if possible have enlisted to the Cause a people full of sympathy for the South. The incursion proved that purpose vain. Eastern Maryland secessionist, yes, including the chief city, Baltimore. Western Maryland, no, except spots that greeted him with enthusiasm.

Lee's grand offensive scared the North but sowed evil for the South. Many Northerners inclined to peace as long as rebellion stayed home. A rash of hatred scared off the peace when the Southern neighbor threatened Northern soil. On September 22, Lincoln could do what had been opposed: issue the preliminary Emancipation Proclamation. As emancipation would free slaves in United States, so it sanctioned slave uprisings in the South. The chasm could not be shut now. With renewed heat on both sides the people girded to continue. The North wanted moderation no more. The general of moderation, McClellan, paid for his hesitant style. Ambrose E. Burnside replaced him.

## 4

Burnside, there to replace moderation, knew as if ordered that he must be bold, the stroke decisive. He tried something harebrained. Lee stood at Fredericksburg, forces concentrated in a position as beautiful as could be, south of the Rappahanock. On December 13, Burnside attacked Lee's front. Lee's Confederates beat back Burnside's Federals and inflicted loss upon loss. Lee never left the defensive, stayed safe, suffered little, and won besides.

The Union, hoping once more to cross the Rappahanock, demonstrated that operations must fail in the American winter. In January 1863 the army tried to go south. Lee knew his homeland's roads. He stayed put. Union columns struggled for days to free themselves from mud: mire without bottom.

Fighting continued in the West. We need mention only Corinth. The three days ended like all western encounters then: a draw. The West, wide open spaces and few people, demanded unity of purpose and clarity of vision across a chunk of continent. Uncoordinated battles and shortsighted leadership brought the Union a strategic nought. On the Mississippi, tactical gains in steady advances along that lifeline, yes, but only there. A block to strategic success had asserted itself—stoutly fortified, stubbornly defended—between boats below and boats above. Would the Union decide the West, it must remove the block. It must smash Vicksburg.

## 5

Burnside, leading in Virginia, had been reckless at Fredericksburg. The Union chose his replacement from among those who had censured him, none other than General Joseph Hooker. His friends called him Fighting Joe.[11] In 1863 his Army of the Potomac numbered

160,000, the largest it had been yet. Compare Lee's 50,000. And then Lee, because of a bad harvest, had to scatter them after provisions over the region. Against Union advances on the peninsula he must detach D. H. Hill to strike back. Worse, the cavalry could muster but two weak brigades, having had to reorganize after sickness when raiding last into Maryland in 1862. Lee's headquarters dreaded the future.

A frontal assault had been the end of Burnside. His successor must try the flanks. Hooker would feint to the front to hold Lee, meanwhile take the main force over the Rapidan. Whence either hit Lee in the flank or advance on Richmond and force him to abandon his redoubt.[12] Lee—cavalry ever active, intelligence system keen—learned the intentions in time. He hastened scattered units together, left a shadow in the redoubt, and met his enemy in the Wilderness, a region of forests. Hooker, heading a mass hard to maneuver, could not array it for battle in time. Jackson surprised him and smashed and annihilated his right. Federals had fortified some positions, but Jackson's stroke enabled the Confederates to unite and the next day (May 3) drive them out. The Confederates won at Chancellorsville against three times their numbers.

General John Sedgwick also failed in the feint at Fredericksburg. He pursued his opponent, Hill, withdrawing toward Lee into thickets south of the Rappahanock. Lee, chasing Hooker retreating into a second position, took messages, heard the cannons' thunder grow, and knew the score. Quitting Hooker, he hit the amazed Sedgwick the next day with two divisions. The losses drove Sedgwick back over the Rappahanock at Banks's Ford. Lee returned with forces disposed to pound Hooker's second position but found nobody home. The river at the camp's rear had risen to a threat of a flood. Ahead of it the Union army slipped out the night before.

Again a general-in-charge had proved incompetent. Again the North lay helpless, open to Southern attack. Yet, against gains, the South must reckon a loss heavy enough to tilt the balance almost to negative. Jackson, wounded by his own men in the Wilderness, had died. The drive, the force, the courage behind every endeavor had gone to the grave. The emptiness could not be filled. The victory must be at best Pyrrhic. Immediately the Confederacy felt the loss. For sure the Confederacy suffered in the next offensive: the invasion of Pennsylvania.

## 6

The Confederate government, with President Davis's energy, had decided to invade once more. They wanted to replenish stores and relieve Virginia of war's ravages. And should fortune favor the Stars and Bars, they would capture a northern city and (with it as hostage) from it dictate peace favorable to the South.

Before Lee headed north, he tried trickery and deception to keep his purpose from Hooker. Hence, among other things, Lee's reconnaissances and feints: the one diligent, the other deft. Feints caused the cavalry battle at Brandy Station.[13] Lee used the time meanwhile to assemble from every part of the South the troops that could be spared, seventy to eighty thousand. Then he described an arc below and west of Washington, and went down the Shenandoah valley and across the Potomac at Williamsport, Maryland. By the end of June, via Chambersburg, left flank for a moment up as far as Harrisburg, he arrived at a certain Pennsylvania town. He protected his right with cavalry during the trek. The cavalry plugged the gaps in the Blue Ridge of Virginia, fighting Union cavalry in the stiff engagements at Aldie, Middleburg, and Upperville. The enemy meanwhile, under General George G. Meade, described a lesser arc between him and Washington. At Gettysburg they collided. Lee's cavalry, under Stuart, ought to have shielded the front and flanks of the army, but, on a useless raid, it sneaked behind Meade. Lee lost what he had always enjoyed: the antennae of cavalry that sensed everything before a decisive battle. He knew neither Meade's strength nor site. The blind Lee, thinking Cemetery Hill poorly defended, struck with units spread rather than concentrated. Nature herself had left the hill a redoubt. Meade massed there. In the gore of battle, Lee squandered and wasted every last power of attack.

The Battle of Gettysburg—July 1, 2, and 3—spilled as much blood as any of this war. Lee pursued a grand design. He threw his strength into the offensive and care to the winds. He spent himself, realized it, and must desist, outcome unsure.

He backed to the Potomac by the fifth, as decisive as only he could be, even in retreat. The enemy, shaken and unnerved, followed but with diffidence. Lee, partly of need to bridge rising waters on the eleventh, offered battle once more north of the river. He chose an excellent site and fortified it fast. The enemy kept clear, as if in the wisdom of detecting a Trojan horse. Lee's army reached the home shore unmolested.

Later, trying to shake the enemy, Lee dared at the enemy's flank.[14] The enemy had improved in a year, hapless Pope no longer commanded, and Lee's thrust gained nought. Then Meade crossed the Rappahanock, took the Confederates' outermost fortifications, and gestured as if southbound. This advance, equally daring, also gained nought. Winter neared. The war in Virginia paused, taking the winter recess.

Both sides understood the summer's news from the West. Grant, attacking at last from landward, on July 4 took Vicksburg. Many western states had supplied man after man and matériel upon matériel to the South. Now the prodigals lay crippled and bound and the West open to river invasion from the North. The news produced joy on one side, dismay on the other.

In his western campaigns, and especially in that siege, Grant demonstrated ruthless energy; tenacity consumed him; he manifested military talent. Lee the undefeated? Northerners had reason to believe his match had appeared. Grant, named general-in-chief, took command. The Army of the Potomac and all the Union armies would take orders from him now. Equal caliber would be trained on Lee. He would meet his match.

## 7

The siege of Charleston occurred at this time, 1863, but meant nothing to the strategy of the war at large. In what the North regarded as the seat of rebellion, the Stars and Stripes had suffered initial scorn. There the national honor had been sullied. The siege, though irrelevant to the rest of the war, would use the power of navy, artillery, and engineers to crush a symbol. Indeed, the siege's purpose amounted to a Union display of its superiority in navy, artillery, and engineering.

A map will show the situation. The offensive must be by sea; swamps guarded the city landward. In April 1863, Admiral Samuel F. DuPont's armored gunboats tried to surprise but failed to take it. In July—General of Engineers Quincy A. Gillmore heading the attack, Beauregard the defense—a combined operation began: engineers and the navy versus harbor fortifications: Gillmore and Admiral John A. Dahlgren attacking in concert and coordinated. Gillmore landed on Folly Island, went to the narrow, sandy Morris Island, zigzagged and dug parallels against Battery Wagner, and at the same time opened a bombardment of Fort Sumter. The night of September 6/7, on their own initiative, the defenders left Wagner. Gillmore reduced Sumter to rubble, tried once more but failed to take it. So ended the expedition, expense high, success slight, nothing to show but the capture of Morris Island.

## 8

The spring of 1864 saw the Northern armies swollen as hitherto unknown. Grant, directing all as general-in-chief, massed and took personal command of superior forces along the Potomac.

### How superior?

| | |
|---|---|
| Army of the Potomac, General Meade | 194,700 |
| Army of the James, General Butler | 69,100 |
| Army of the Shenandoah, General Sigel | 30,800 |
| Cooperating in Virginia (total) | 294,600 |

### How inferior?

| | |
|---|---|
| Lee against Meade | 52,600 |
| Associated forces: Richmond garrison, General Robert Ransom | 5,000 |
| Troops in North Carolina, Generals George E. Pickett and Robert F. Hoke | 10,000 |
| Wilmington garrison, General William H. Whiting | 7,000 |
| Charleston detachments, pickets, etc. | 3,000 |
| Other small detachments | 3,000 |
| (total) | 80,600 |

Thus Lee's *summa summarum:* no more than 81,000 to defy about four times the number. To judge the rest of the war, bear the proportions in mind.

Consider further a Northern infantry better in tactics now, and stricter in discipline, under generals who had learned to command; adequate staffs directing movements; and a cavalry escort big enough and in tip-top shape. Can we but conclude that the Confederacy awaited battle with foreboding? Apprehension grew the more as the

North enjoyed daily such an influx of manpower that whatever the defeat, losses could be replaced out of huge and untouched reserves. The Confederacy had not only armed its last man, it had also, owing to the misfortune in the West, seen four of its best provinces cut off, those sources of men and matériel lost. Advantages in manpower thus crowded upon gains in matériel in the North. The South must assert the genius of leadership and the courage of men in arms—it had nothing else.

Grant crossed the Rapidan, like Hooker the year before, to lure Lee out of position south of the Rappahanock. Lee, informed by alert cavalry as before, and doing as he had, moved at once to meet the attacker in the Wilderness. Here the thickets a year before to the day had seen gore; now the first gore of Virginia in 1864 turned them crimson. Two days ended in a draw because Longstreet's men shot their commander at the start of the decisive thrust, like Jackson's the year before.[15]

Hooker had stood fast. Not Grant. This man of energy, hurrying to finish Lee, turned left and tried between Fredericksburg and Richmond to slip behind him and cut him off from his capital. Grant's flanking column arrived at Spotsylvania Courthouse to find Lee waiting. Grant had decided to hammer resistance to bits bit by bit. With tenacity seldom seen anytime, and strength rare anywhere, he took on the Confederates then and there. The struggle, more of the shovel than of the gun, lasted eleven long days. The gun counted too, and drew blood.

The Battle of Spotsylvania ended and preliminaries to the Battle of the North Anna began on June 19.[16] Lee saw Grant reach with his left, stretching and straining, trying again to force him away from Richmond. Lee therefore moved his own left out of entrenchments. The feint hit Grant fast and hard. Lee's main sneaked meanwhile to Hanover Station and dug in on the other side of the North Anna River. Grant had intended a quick move to outflank him south. In Lee, however, he faced an enemy whose cavalry kept him informed and who divined everything besides. Surprise joined Grant's confusion when Grant, already perplexed by Lee's feinting, again found Lee opposing him and from a site well chosen. Grant dared no serious attack. He undertook an energetic, intensive reconnaissance and tried to keep Lee in place. Then—and striving for secrecy—Grant followed the Pamunkey down to Hanover Town and crossed. Would he manage to get around Lee at last? He found Lee on the upper Chickahominy, waiting. Grant had lost touch; he could no longer communicate; every line had broken. He went south to the York, set up a supply center at White House, and relied on the fleet for distribution. He had fought the war's bloodiest battles so far and fetched up where McClellan started two years before.

Jackson had hit McClellan at Cold Harbor, back then. Now, having brought Union forces to Cold Harbor again, Grant assessed his opponents as weak and intended to smash them once and for all. On July 3, 1864, he hit Lee straight ahead: a frontal and nothing but a frontal. Without stealth or deception, with nothing of the craft to reduce resistance and aid the assault—no flanking, no feinting—he hurled all he could; he flung everything headlong and headless into battle. The brief, savage onset ended in colossal, bloody defeat.[17] It proved to Grant that power without craft would not outdo Lee's veterans. Indeed in the summer's battles—can they be named in their confusion and counted in their multitude?—Grant lost something like sixty thousand men in the carnage. Lee perhaps a mere eighteen thousand. I say "mere" to stress the comparison. To Lee, however, comparative success could not offset absolute loss. With him every loss must be absolute. The North could replace Grant's. Lee must do with what remained after losses. His army dwindled and shrank.

Subsidiary and associated forces had been active. Strengths resembled those of the principals. Here, too, speed and energy might prevail; the South relied on them against numbers. When Generals Franz Sigel and John C. Breckenridge clashed in the Shenandoah in mid-May of 1864, Sigel of the numbers lost his command, so badly did he do against energy and speed. Breckinridge, forces nearly intact, went from there to fight beside Lee at Cold Harbor. Below Richmond, on the James, in May and June, Beauregard turned Butler, a slice so sharp that Butler gave up the offensive and dug in at the Bermuda Hundred.

Grant, trying to break the ring around Richmond, watched his tries from the north fail. He would try south. He circled Richmond. By every effort, especially by taking Petersburg fast, he would get the key to Richmond as well as seize the terminus of railways south and west, the Weldon and the Lynchburg routes, and cut off Lee's supplies. Only eight thousand held Petersburg in fortifications styled after the tactics of the time. The weak complement, under the cleverest of leadership by Beauregard, stood against the strong enemy long enough for Lee to arrive in force. Again the Union paid, this time with fifteen thousand veterans. Fresh troops stepped in but, green, could not restore an army shaken by defeats. Grant's try after stubborn try against Lee, by onslaught after bloody onslaught, had failed.

Grant invested Petersburg in mid-June, fortified himself by extended entrenchments, and addressed Lee as if for the usual attack. None had started like this. By the end of July a mine had been dug to put eight thousand pounds of black powder under the Confederates. It exploded. Attack followed. The Confederates beat it back. After such ferocity the war sought the peace of the trenches. Union detachments, out to cut the railroad west to Lynchburg, found trackless forests and got lost. Assiduous Confederates chased them off without trouble. Nothing remained to Grant but a steadfast, stubborn campaign by pick and shovel.[18] It too would have gone nowhere, only the war in the West took a turn fatal to the South. Sherman began to march east.

## 9

It had been ebb and flow in the West until the spring of 1864. Turning back the expedition on the northern Red, the Confederates cleansed the far West of Yankees and recouped in part the disgrace at Vicksburg. Winning at Chickamauga in Tennessee, the Confederates regained a measure of the Chattanooga loss. The Union held Knoxville but could show nothing else for endless fighting there. In sum, battles and movements occurred one place and another, and diverse and sundry operations dissipated superior numbers, until Grant consolidated the numbers and focused the action. He named Sherman commander-in-chief west, and merged under him the three armies: Tennessee, Ohio, and Cumberland: 200 short of 100,000 men. Against them, Johnston could muster 56,000 plus 4,000 cavalry.

Defense has advantages, however. Johnston availed himself of them. Sherman met the skill of Johnston on defense: Sherman on offense ever moving to flank him on the left. So they seesawed. The tilt favored Sherman: Johnston lost terrain: again and again. His president and press believed an offensive Virginia-style from time to time must prove effective. General John Bell Hood therefore replaced him, July 17, 1864. Hood with an army to equal Sherman's in morale, unequal in numbers, could not manage the offensive, Virginia or any other style. He tried. Sherman with maximum effect parried those brilliant thrusts.

Sherman's genius won. Taking Atlanta, he weighed the South's anchor to operations in the West. Worse, losing Atlanta, the South lost the center of the supply system essential to resistance. Sherman had run the South through to the heart. Devastating for the South, nobody opposed him between Atlanta and the sea.

Before him lay territory rich, fertile, and untouched by war. There he could move and be sure of provisions; he needed no supply lines. So it occurred to him to cross straight to the Atlantic. The expedition and its unfortunate devastation of Georgia cut Confederate lifelines, of course. Savannah evacuated when he appeared, Charleston too, and he marched north. Lee would be isolated in Richmond; he must fear encirclement with nowhere to go. The end neared.

## 10

Lee accordingly, not wanting to be smothered and crushed at Richmond, abandoned its fortifications. He would take Lynchburg if he could, join Johnston, and fight a decisive battle in a region not yet bled and depleted. He ordered provisions to Amelia Courthouse. Misdirected, they went elsewhere, leaving his troops not only driven to exhaustion but pushed to starvation. Many scattered, trying to find food somewhere, anywhere. The enemy surrounded him. Still he defended himself like a lion against every attempt on his pride. The way to Lynchburg blocked, his men too tired, too hungry to march, only then did he surrender himself and eight thousand to an enemy ten times the size. His career ended as it had begun: heroically. The other scraps of Confederate armies followed, sealing the Confederacy's fate in April of 1865.

# 10

# WAR IN THE WEST: COMBINED OPERATIONS

*The river meant everything to victory in the West.*

Combined Operations Defined—Along the Mississippi—June 1861 to May 1862: Fort Henry to Island Number 10—May 1862 to July 1863: Siege and Capture of Vicksburg—The End (July 1863): Capture of Port Hudson—Consequences, Observations, Lessons.

**S**hips and boats may be inferior to railroads in warfare. Navigable streams do not run through mountains, for instance. Nor do ships and boats roar across the sierras. Nor can a network of waterways be built over a vast area at low cost. Waterborne transportation, master of the high seas, goes slower than the railroads across the plains, creeps through the piedmont, and stops at the fall line. Yet ships and boats in arms can prove superior to railroads. There need be only navigable waterways or the sea itself at hand. With those words I have said enough of what every schoolboy knows. Let me leave the obvious for truths less familiar but more valuable and worth greater attention.

One by one:

(1) Ships and boats must keep to their routes, true, but these cannot be so easily destroyed as railroads. Many a time the patrol of cavalry and the stick of dynamite suffice to interrupt rails when and where the patrol chooses. To block a waterway demands larger means and much engineering—if it can be blocked at all. Many cannot be. This fundamental advantage counts all the more in these times of large, extended operations and fast, easy movement.

(2) Ships, boats, and ancillary craft, hold more than trains. Troops —and everything they need of provisions, arms, munitions, equipment, and the rest—can be loaded together and in numbers and bulk

until a small flotilla carries a large force. Railroads would have to deliver them piecemeal. How significant the difference? We see it—it counts—when troops go from vehicle to combat, as in Asia and Africa and countless times in the American Civil War.

(3) Waterborne transport does not put the railroad's stress and strain on passengers. They lie down, move about, and satisfy their every need. Neither tired by the railroad's cramped spaces, nor unstrung by rattle and vibration, troops reach combat rested, fresh, and ready to fight.

(4) Troop ships can be armored against enemy fire. Moreover, armed with cannon and manned by sharpshooters, they assert their own firepower; the ships themselves can be decisive in battle apart from their troops. Of course such effectiveness stops at the range of heavy guns. Still the value of ships in combat ought never be underestimated. Where for instance do we find installations of military significance, put there of necessity and by design to fill military needs? Need I say on waterways?

(5) Our waterborne vehicles offer another and by no means lesser advantage. They store provisions and ammunition. Given ships' capacity, together with communication secure on the water, an army can be surer of support than if served by the railroad. Unlike ships, the railroad must have stations and the elaborate supplies and complicated facilities endemic to them. Movement then depends more or less upon fixed points. A modicum of ships however, what with their mobility and volume, give you a shoal of independent, moveable warehouses. They accompany the troops and, conditions permitting, assure provisions and other needs. To storeships can be attached self-sustaining combat vessels: gunboats, mortar barges, and landing craft with infantry and cannon. The squadron becomes a floating quartermaster corps complete with defense and ready to defend themselves. They can beat off all but the overpowering attack.

I have thus enumerated the advantages of ships and boats to warfare around ample water. The navy therefore ought never be undervalued. Whatever their advantages, however, ships and boats cannot by themselves achieve strategic objectives. The navy means little to comprehensive strategy except together with the army in combined operations. In these pages for that reason I address those operations: the cooperation between ground and naval units, and especially the role of amphibious forces. Conditions permitting, a nation will improve its position by articulating the army and navy, thus marrying offensive power to defensive strength.

Our discussion ought not remain abstract; it needs illustration by events, an example from experience. I have chosen the American war's Mississippi theater. Even the tributaries there, though not as deep as our larger rivers of central Europe, could float lesser-draft steamers and gunboats as well as smaller craft. So the Mississippi system figured large. The Americans assembled an inland flotilla of significance, however improvised. The example will teach us how a naval arm can make the difference in aid of scouting and foraging in addition to the capture of territory and assisting in its defense. Put the navy hand in glove with the army, let the flotilla take orders from the commanding general, and you cannot overestimate the value of the partnership in transporting troops, forwarding supplies, and in the fighting itself. Consider Foote's cooperation under Grant, say, and Porter's under Sherman.[1]

## 2

The states had been United. Civil war divided them into North and South, Yankee and Rebel, Union and Secession, Federal and Confederate. Americans on both sides used those names. The antagonists fought a bloody war four years long, until the states of separatist intent had been crushed and the United States restored. Here and there along the coasts, at the mouths of rivers, and elsewhere, dilatory engagements took place. But major actions in two theaters, Virginia and the Mississippi valley, settled the issue. What happened there determined the outcome everywhere.

Those theaters and the locus of fighting in them could not have been more different, East and West. Fighting in the East, in Virginia, mostly went back and forth above and below the Rappahannock—a nook in the American states, a dot on the continent. In the West the fighting burned a line between St. Louis and the Gulf—miles that would reach from the Mediterranean to the North Sea.

The presence and absence of a navy mattered. The industrial North had the navy, but it could not operate in the eastern theater. Fleets of fighting ships could not negotiate even the major rivers: Potomac, James, and Rappahannock. Nor did control of the coasts matter; a wilderness of swamps divorced sea from solid ground.

McClellan tried to go "on to Richmond" through the swamps of the Chickahominy but came to grief. I can think of no incident better to illustrate the maxim: Don't campaign in swamps. The absence of the navy left the mediocre army to face one that enjoyed advantages beyond measure: the best generals, Lee and Jackson; the best soldiers, Virginians; and the best area for their fighting, Virginia.

How different the West! The great river and its deep and powerful tributaries in the middle of the country connect states to one another and all to the ocean. This Mississippi, purveyor to a lion's share of the continent, constitutes transportation, communication, political no less than economic power, and life itself. The Indians accordingly settled the shores first. Today the currents impart vitality and carry the loads to city after city along the shores. No wonder, when civil war seized and shook the nation, the already grand waterway gained significance. Control it: control the West. Control the West: hold the strategic upper hand. But to control would take two years of war. Before we address the war there, let us look at the complications posed by geography.

The basin spreads more than three million square kilometers. The Allegheny Mountains stop the reach to the Atlantic. The Nevada and other ranges bound it on the west.[2] In between, a network covers and interconnects most of central North America: the world-class river system of the Mississippi plus feeders, notably the Arkansas, Black, Ohio, Tennessee, Cumberland, Yazoo, and Red, together with the bayous along the awkward, twisted, tortuous lower course. Tributaries and bayous, navigable and posing strategic and tactical problems, figure in the war and interest us.

The Mississippi itself, navigable to St. Anthony's Falls in Minnesota at 45 degrees north, empties into the Gulf below 29 degrees: enough to reach across most of western Europe north to south. Miles and miles north, headed south, the Mississippi makes the promise not kept until St. Louis: to be the large and the significant, the continental bearer of commerce. At St. Louis, fed by the abundant Missouri, the Mississippi grows giant and rolls the yellow flood through level countryside known for fertility and renowned for bayous, to the Delta's broad forks and beyond. Beginning at St. Louis the Mississippi therefore, at average levels, can float big boats unimpeded to the Delta. Western tributaries dump into the southern river the masses of trees uprooted at high water, but the trouble to shipping never amounts to a break in communications. Our concern begins at St. Louis and runs to the Gulf.

Opposing armies of about equal quality could not find a place for a decisive battle in this large yet confining terrain. Nor did a second Lee appear with his generalship of genius to compensate for a want of other sinews of war. The student of this struggle, paying due attention, must conclude one thing and no other about how long it lasted and the way it ended. I mean that the war here would have spun out to unforeseeable length, with who knows what outcome, had not the navy played a starring role and often the lead. Quickly the Union assembled a river-going fleet. It found the Father of Waters and his family hospitable, their waters a good place to fight. By virtue of that network, the fleet encircled the Confederate army, interdicted its supplies, drew tighter and closer to impair and reduce its effectiveness, and cripple it, and hack it to pieces.

That Union fleet, partly combat ships, partly transports, guaranteed everything to the forces ashore: provisions, ammunition, replacements, and back-up fire. The Union could move armies with ease hundreds of miles in one direction and another, without tiring the men or putting them to the stress of bad roads and other hardships of movement overland. The fleet could also join battles against artillery ashore. Naval artillery more than took part: it *mattered*. The fleet brought to battles near rivers all the more firepower because gunboats floated in so easily with ordnance bigger than the field artillery the enemy had to drag in on bad roads. Nothing but coast artillery and heavy batteries of a few forts could do extended and serious battle against most armed and armored rivercraft. Ships damaged could usually withdraw: another advantage. Soon repaired and resupplied out of danger, they would return reinvigorated, strength restored, and rejoin the attack. Applying the navy's effectiveness to combined operations, the Union won the West.

3

History probably offers few scenes of public strife and personal crisis so dramatic as the western United States in 1861. When civil war broke out like a rage of the elements and a storm upon the land, the jolt shook society to its foundations, unsettling even family and home. The war it seemed would occur mostly in the border states because supporters of North and South lived nearest one another there and mingled one with another and some with others. Thus the rupture and the distress, holding this brother home and driving that brother to join the enemy, must be sharpest and deepest in Kentucky, Tennessee, Missouri, and Arkansas. (Not to mention Maryland in the East.) Rifts opened, not only in government, commerce, and public life, but even between fathers and sons, and among brothers and sisters. Hence the savage local engagements and the confusion in the West when war broke out. One side succeeded here, the other there, between June and the end of 1861. Circumstances did not clarify enough to reveal tendencies until winter, 1861/1862. We cannot speak of patterns and trends until then. At any rate it would exceed our purposes to mention the many and bloody skirmishes. Let us therefore begin with a summary of the situation at that time.

Union forces under Grant—holding Cairo, Birds Point, and Cape Girardeau—had fortified and secured those keys to the Mississippi. Thus the Union controlled points of strategic significance: the mouth of the Ohio as well as of the Ohio's tributaries, the Tennessee and the Cumberland. Grant's men had thereby cut off from the Mississippi the Confederate steamboats on the Cumberland. In addition, Kentucky belonged to the Union now. In the meantime the secessionists under General Leonidas Polk had secured Columbus in Kentucky, New Madrid in Missouri, and Island Number 10, Fort Pillow, and little Fort Randolph in Tennessee. Polk's men thus had interdicted Mississippi traffic. At the same time the Confederacy drew a line from Columbus through Forts Henry and Donelson in Tennessee, to Bowling Green, Kentucky: the line of defense. Union forces had settled in along the Ohio, to threaten the Confederacy from the north along that line.

Now we can see the Union's plight. To move its forces of men and boats deeper into the Confederacy the Union, having secured the mouths, must gain the lengths of the Tennessee and the Cumberland. The Confederacy controlled them by holding Fort Henry on the Tennessee and Fort Donelson on the Cumberland. Wanting to move boats as well as men and matériel, the Union would attack by land and by water.

At Henry, General Lloyd Tilghman commanded 3,400 men and 17 guns. The site—it looked up at the bluffs of the Tennessee rising all around—could claim little as a works except for the shape:

five-cornered. On January 30, Grant headed seventeen thousand for it by water aboard the flotilla under Commodore Andrew H. Foote. The commodore had divided his forces. (1) Four gunboats: *Cincinnati* (command boat), *Carondelet* and *St. Louis* (thirteen guns each), and *Essex* (nine). (2) Three wooden boats: *Conestoga, Tyler* and *Lexington*.[3] On February 4, Grant's troops began arriving and landing at Bailey's Ferry. Tilghman—before the landing ended on the sixth—assessed his fort's situation (wretched), decided to defend it as long as he could with one company of artillery, and sent his infantry under Colonel Adolphus Heiman the fifteen miles to Fort Donelson.

Foote's first division, beginning the attack on the sixth, opened fire from 1,500 meters. Slowly, firing forward guns only, they approached to 500, closed in, swung broadside, and let go with everything at the fort's batteries. Meanwhile, standing off, the second dropped shells into the fort.[4] The handful of defenders, returning a fire to be noted for its vigor, scored 59 hits. All but 1 bounded in vain off the armor. That 1 counted. It split the boiler of the *Essex*, leaving 28 men and the captain, William D. Porter, wounded. In the fort a rifled 24-pounder burst and disabled the crew. A 24-pound missile exploded at another gun: 3 killed, 1 wounded. Misfortune joined injury when a jam rendered a 10-inch Columbiad useless. Add the other 5 killed, 11 wounded, and 4 more guns demolished. In sum the defense had been reduced until Tilghman himself served a gun. Meanwhile he did all he could to restore morale to men disheartened by havoc within their walls. Chances for success shrank so much that officers not on the firing line abandoned the fort. Tilghman, every resource exhausted, had 66 men and nothing else. He waved the white flag to overwhelming forces.

Fort secured, the flotilla split into its divisions: the first to Cairo, the second to operations on the Tennessee. North and South turned attention to the next concern: the one to attack, the other to defend the fort that guarded the Cumberland.

Donelson surpassed little Henry in size, location, weapons, and fortifications. (A map here would be no more than as sketchy as those of most Union engineers. Theirs, supposedly drawn on the spot, betray ignorance even of basic geography.) Just below the fort two shore batteries had been set up at about ten meters higher than water level: a ten-inch Columbiad and nine thirty-two-pounders in emplacements and capable of a wide field of fire, in the one nearer the fort; and a forty-pounder and two thirty-two-pounders in the other; all smoothbored except the forty-pounder; all firing through slits, target area being small. In the fort itself: eight cannon. The infantry and its light field artillery took defensive posts behind scanty earthworks from Hickman Creek to the town of Dover. Thus in a semicircle along the riverbank the infantry had spotted its ordnance to advantage along the front lines. By now the defenders, increasing slowly, numbered about fourteen thousand.[5] John B. Floyd„ the senior general, took command. Generals Simon B. Buckner and Gideon J. Pillow headed the right and left wings, respectively. Leadership remained unclear, however, with Floyd not one to secure obedience. Indecision and incompetence ran rampant, leaving things all the more ominous because troops dribbled and trickled in late. Little time remained even for a semblance of organized defense.

The moment Grant took Henry he decided to attack Donelson. He sent fifteen thousand men overland despite roads sodden and flooded. On December 11 he had communicated with Foote about cooperation. Foote declared naval forces ready now.[6] Steamboats brought six regiments to the new battlefield a short while ahead of the fifteen thousand. On February 13 the two contingents mounted a surprise attack. Of naval forces, only the gunboat *Carondelet* could join. Standing away, she engaged the shore. The upper battery's forty-pounder reached her at that range and sent a projectile through a port. It damaged the machinery and she limped off. The rest of the attack failed, too, casualties severe. The attackers postponed operations and waited for the navy and reinforcements, using the time to organize the shoreside assault. Its details do not belong in these pages.

The rest of the naval forces arrived on February 13 and 14 and by 3:00 P.M. had begun to attack. Command boat *St. Louis* and the three other ironclads crept forward, leading two timberclads. The flotilla opened fire, range about 1,500 meters. The enemy replied. They continued: the attackers approaching and firing, their opponents returning the fire. Arrived in the face of bombardment to about 400 meters, the attackers began firing forward guns, three in the bow of each boat. After but a half hour, 2 ironclads had been disabled: *Louisville* (damage to paddle wheel) and *St. Louis* (steering mechanism). One of the *Carondelet*'s rifled cannon had blown up. The flotilla must withdraw, together with 2 other boats also damaged and

54 men wounded, including Foote. The flotilla had been aiming at the earthworks but done no harm.

Yet on the sixteenth the fort surrendered—under incredible circumstances. With so many troops crowded in, the fort had become like a straitjacket, constraining the troops, making more defenders less effective, and weakening the defense. Constriction and congestion demanded that as many as possible be got out. On the fifteenth the necessary sortie succeeded, reaching its goal, Wynn's Road, when the order came—for reasons unknown and unjustified—to return to the fort and do something even more disgraceful: give up to an enemy practically no bigger than itself. The fort surrendered after Floyd as well as the cavalry, some of the infantry, and several boatloads of wounded escaped.[7]

New Madrid and Island Number 10, well situated and strong enough, interdicted shipping, no doubt about it. Before the river could be of use to the Union, their armament had to be knocked out and the battery smashed that commanded the river from the right bank south of Number 10. On February 14, General John Pope got the order to attack. At the end of February the Union ended the defense of Columbus, Kentucky, letting its troops be sent elsewhere. Pope divided them between New Madrid and Number 10, and sent the artillery to the battery south. We need not describe the topography; a map will show the essentials for us: two horseshoe bends, New Madrid in one, Number 10 in the other, and both in marshy terrain. Here we need note only New Madrid's lack of elevation and weak walls. Walls? Within mere breastworks thrown up in haste and neither high nor thick, the fort hunkered so near river-level that at high water, Union gunboats could not only rule the terrain outside but also fire over them into the fort itself.

Could Pope mass the force to storm it? In a region of flood upon flood, swamps prevailed. They seemed to say: You can't walk through here; bring up the blasters. Siege artillery arrived on boats. The defenders' ordnance, not redoubtable for power, had to be feared for the fix and repeated hits on gunports. The boats must stand off. At last they anchored along the opposite shore.

The attackers managed to start to build approaches. They would not be necessary; siege lasted only a few days as low morale overpowered rank and file of the defense. Without their officers' permission they began to evacuate, three thousand the night of March 13/14, emptying the fort. They got across the river and to Number 10, and brought along the artillery that could be moved. But as, lacking discipline, they lost their nerve; so, lacking discipline, they neglected the essentials. Therefore they abandoned to Pope an abundance of cannon and other matériel.

No sooner had Confederates left than General Schuyler Hamilton's division moved in, rearmed the works with captured cannon, and commanded the river. Still the Confederates could approach from downstream. So at suitable places on the right bank south to Point Pleasant, Union batteries joined this guarding of the river, also with captured cannon. The Confederates could have approached before Pope cut them off. Now he controlled the river south to a point opposite Tiptonville. On the fifteenth the Confederates discovered the new batteries. Five Confederate gunboats came within 300 meters, opened fire, and poured it on. After only an hour and a half, however, one had been sunk, the rest damaged, and many cannoneers killed by sharpshooters. The gunboats had to withdraw.

Foote the same day brought his mortar-and-gunboat group to the vicinity of Number 10. To reconnoiter defenses and learn how many and how big the artillery, he engaged at long range the island's emplacements and those ashore as well. Next day, 4 mortar and 5 gunboats neared to about 1,800 meters. Opposite the upper batteries they opened fire. Shore and island responded. More than 100 guns thundered. The breastworks, 8 meters thick but on a soft basis and wave-drenched at the escarpment, gave way under bombardment. Thirteen-inch projectiles exploded in the breastworks. Through the holes came the deluge. Defenders waded in water and mud. Shelling demolished cannon. Another blew up and killed several. On the *Pittsburgh* an attacker's did the same, disabling 14. The Federals continued the attack anyhow, but the Confederates neither gave in nor gave up. Even after several days' battle their guns would not be stilled. Those at the fort (not those ashore) shut up later. Silence demanded the landing of a detachment, the overpowering of the fort, and the spiking of the guns. But Union forces could not get established ashore.

Meanwhile (March 17) General John P. McCown's Confederates assembled six regiments of infantry, some cavalry, and two batteries at Fort Pillow. They went to Corinth. General A. S. Johnston put them into the larger force he assembled there. They marched to the river town of Tiptonville. Johnston arrived on the twentieth and took them to the peninsula opposite New Madrid.

General Henry Halleck, Grant's superior in the West then, had advised Pope to build a road along the right bank, so be able to site batteries to support the navy against the artillery on Number 10 and artillery nearby ashore. Reconnaissance said a road could not be built nor batteries placed. "But the island can be flanked. Dig a canal through wetlands and swamps to Wilson's Bayou." Dredges, small boats, light steamers, and other equipment set to deepening those shallows and dredging a channel. A mass of old, sunken tree trunks impeded progress. Some had to be sawed, others pulled by machines. Trees to the right and left, trailing foliage and branches across and into the canal, had to be trimmed or removed. Men often worked in water up to the chest, one of the ordeals, yet completed the toil in nineteen days in a canal wide and deep enough for light steamers and small boats. Several started through on April 5.

McCown, commander of the Confederate forces, and his successor as head of operations in this region, General William W. Mackall, had heard of the project; it had been reported to them. They dismissed it. "A canal can't be dug there." Neither acted to stop what he thought impossible: neither took measures against its consequences. Captain of engineers A. B. Gray—he first of all—reported it completed. He—only he—acted. To secure Confederate positions he put batteries near New Madrid to bombard the exit.[8] Would their fire block it? Steamers and small boats, the Union's test flotilla, started through. Flanking the peninsula, they neared the exit. Fired upon, they could not get out.

On a dark night, Pope said to Foote, move two heavy gunboats past Number 10's batteries, and on toward New Madrid, and knock out those interdicting the canal's exit. Too dangerous, Foote said.

Pope decided, Foote notwithstanding, to create his own floating batteries. Assemblies of barges or other boats would carry the ordnance. Assemblies would be of three substantial craft bound into one, and three cannon mounted in the middle craft. As traverses, four feet of planks reinforced by sandbags would protect the cannon. Bales of cotton stacked on the outer craft would further stiffen the shield. Planks plus bags plus bales added up to breastworks twenty feet thick. Empty barrels fastened to the hulls would aid flotation should enemy gunfire punch holes in the hull. These floating fortresses would weigh anchor after dark, ride the currents past the enemy's cannon, to the spot marked for landing, put the troops ashore, and open fire and rain destruction on everything intended to stop them.

Did it happen so? Not at all. In the mean time, Foote said yes when the brave Captain Henry Walke volunteered to take his heavily armed *Carondelet* past Number 10 and on to New Madrid.

Boats would be readied and batteries run again and again during this campaign. We should therefore pause to describe the first, by Walke's *Carondelet*. Bales of cotton ringed the deck. Chains, thick links secured where they mattered, guarded vulnerable spots and especially the pilothouse wrapped in an anchor chain's massive links. A web of hawsers, ropes, and cables shielded the boilers. Guns pulled in, ports plugged, and crew armed would prevent boarding. Still the enemy might overpower her, so the outfitters made sure she could be scuttled.

Night fell.

Twenty sharpshooters came aboard. A coal barge, full of baled hay wet, went to the side to face the enemy, fastened there against his fire. Signs in the skies announced a thunderstorm brewing.

The *Carondelet* cast off, preparations done. Downstream she went, under power but at a crawl. Steam, usually vented in the stacks, had been routed to the wheelbox to reduce noise. Steam in the stacks would have moistened the soot. Now after scarcely a kilometer, dry, it began to burn in the stacks. Fire wrote the message upon the skies. *Look. A boat going downriver.* Ashore, sentries with their small arms alerted fellow Confederates in the redoubts. They joined the defense and sent up rockets and flares to notify the batteries. They opened up and poured it on. The thunderstorm, brewing and brewing, broke now, adding thunder and lightning: thunder above to the thunder below, violence in the heavens to the violence on Earth: one rage upon another. Flames pouring out of the stacks did one thing right. They illuminated the banks in a bright, lurid light. A safe course could be set by it and the order given: full ahead. Crowding the right, shaving it, she steamed downstream as fast as she could drag the barge. She must pass enemy positions. She passed them unharmed. She must pass the enemy's floating batteries below Number 10. She passed them unharmed. The signal for success had been prearranged. At New Madrid at midnight—that soon—she signaled friends ashore: success. She had taken her chances and won. Next morning, April 6, Walke took on board part of Pope's staff, continued downriver, silenced enemy batteries opposite Mount Pleasant, and returned to New Madrid. The *Pittsburgh* followed the next night. She ran the batteries the *Carondelet* had run, and joined her there. On the seventh, obeying orders despite fog, the pair destroyed the batteries at Watson.

Four steamboats of troops could pass then, as the campaign for the upper part of the lower Mississippi continued. Pope, hearing of the enemy's withdrawal to Tiptonville, ordered General Eleazer A. Paine to land the reserves there. Paine engaged Mackall amid the swamps and bayous, and so dealt with his forces that the Confederates surrendered to the last man. Those seven regiments had been lost, their positions therefore lost, Island Number 10 consequently lost, and this part of the river accordingly lost. The Confederacy had lost the Mississippi above Vicksburg. The Union controlled the upper river, every yard and every foot if not every inch. The Union controlled the states along the upper river. In short the Union enjoyed more than enough control to use the upper river at will.

What remained to the Confederates? Nothing but a riverside raid now and then, to trouble the Union fleet or harass one of its boats passing by.

## 4

Confederates holding positions on the Mississippi had closed it and its chief tributaries. The Union would reopen them. When the Union took Island Number 10, the blocks had been removed, all but one. To get rid of it and open the river its length the Union must take Vicksburg. In 1862 and 1863 the Union's western armies therefore concentrated on that objective: By every effort, take Vicksburg. Meanwhile the Confederates strengthened it, their last bastion.

The clash that followed, the struggle for the key to the Mississippi, albeit interesting, remains little known in relation to the rest of the war. I shall tell the story to include the details—if not in relation to the rest of the war, at least as they bear upon our theme.

Farragut undertook reconnaissance up the Mississippi soon after the Union captured New Orleans. The river meant everything to victory in the West. Could New Orleans serve as the base? Could operations from it seize and secure the rest? Rumors ran rife, many unverified. So Farragut set off with a task force to see for himself.

His few boats several days out had met no resistance except mayors' protests as they surrendered cities and towns without means to defend themselves: Baton Rouge, for example, on the eighth of May, and Natchez on the twelfth. Rear Admiral Samuel P. Lee, ordered ahead in the *Oneida,* learned that Vicksburg itself still lagged on defense. Vicksburg nonetheless rejected his demands. Farragut arrived after a while. His force included two steam-powered transports carrying fifteen hundred men and two batteries of field artillery under General Thomas Williams. Studying the fortifications and the rest of the enemy's situation, Farragut decided that his force could not overpower Vicksburg. His boat captains even advised him to drop the plan to continue upriver: Don't pass Vicksburg. Without further ado they went back: troops to Baton Rouge, boats to New Orleans.

In April, Beauregard by his own plan had ordered Vicksburg's defenses fitted for three thousand men and forty cannon. Work began but not in earnest until General Martin L. Smith took charge on May 12. Before the enemy appeared in the form of Farragut on the eighteenth, enough had been done at least to ready six batteries—but only six. Artillery scanty, men few, fortifications weak, Smith dreaded a serious assault. He and Vicksburg heaved a sigh of relief when Farragut's boats turned back, mission not accomplished.

Those events alerted the Confederate army's leaders, alarmed them. Hurry, they said. The defenses must be stiffened, the Mississippi's crux strengthened. By the end of June, fifteen thousand more had been ordered there, and General Earl Van Dorn to command the district.

Farragut, back from reconnaissance, read letters waiting in New Orleans: orders from the Navy Department. One missive did not mince words. We control the river as far up as Vicksburg, it said. Now you take it *all.* He had planned and prepared an attack on Mobile. He believed also that unless aided by the army, he could not overpower a place of Vicksburg's strength. But he followed orders. He dropped his designs for Mobile, banished his doubts about Vicksburg, and sent Admiral David D. Porter ahead. Porter arrived at Vicksburg the same day Farragut left Baton Rouge, June 20. Porter commanded 16 mortar boats, Farragut the bigger hardware—7 gunboats, 3 cruisers, 106 guns in all—as well as 3,000 troops and 2 batteries under General Williams. The Confederates had put batteries at Natchez and Grand

Gulf. When Farragut addressed them, the crews abandoned the guns and fled. He reached Vicksburg on June 25.

Williams landed the troops on the left bank, arrayed the artillery's batteries, pressed 1,200 Negroes into service, and set them to digging a canal through the peninsula. The mortarboats anchored for battle. On May 28 [i.e., June] Farragut, with two cruisers and five gunboats, ran Vicksburg's batteries and anchored above the peninsula after a two-hour artillery duel and 15 killed, 45 wounded.

According to his report, he thought he had silenced the enemy's guns. He had hurt none. He had learned one fact that took no reconnaissance to confirm. His forces on the water and Williams's on the ground, weaker than the reinforced garrison, could do nothing decisive against it. His report sums it up. He says he has passed the forts and can pass them as often as asked but maintains that Vicksburg cannot be taken without an army of 1,200 to 1,800 men.[9]

The boats therefore lay idle. Williams toiled at the canal without pause until high water silted in all he had dug out. Soldier and sailor alike longed to be relieved of this hopeless assignment. Meanwhile the *Arkansas* incident vexed the sailors and set teeth on edge throughout the fleet. Fabulous rumors circulated about that Confederate ironclad. Two light Union gunboats went up the Yazoo to learn whether the rumors held any truth. Could the *Arkansas* be that big, that strong, that formidable? Gunboats briefly gone, Union sailors heard a fusillade, then saw them fleeing as fast as they could, the fearsome *Arkansas* hot after them. Not a single Union vessel had steam. But they could shoot. They let go broadside after broadside. Farragut and his men watched with irritation as she ran them and into the shelter of Vicksburg's batteries, safe.

He decided to go home the next night. Passing Vicksburg this time he would destroy the *Arkansas*. Only, in the dark of the night and the confusion of operations in the dark, she could not be identified. Still he had gained an advantage not to be undervalued; he had regrouped below Vicksburg and lost only one man. Then, to a sigh of relief by Farragut and his men, the order came to give up the expedition. He put Williams and the troops ashore at Baton Rouge, left two boats there, and continued down to New Orleans and anchored, June 29.

Union flotilla gone, a mere 22 casualties in the garrison, need I affirm Vicksburg's jubilation? Van Dorn, deciding therefore on an offensive, sent General John C. Breckinridge and 5,000 against

Williams and 5,000 at Baton Rouge. Fever weakened Williams's forces; they had to fight a pitched battle; their casualties included Williams himself, killed; yet they held Baton Rouge. Breckinridge and the Confederates suffered loss upon loss among officers yet stayed strong enough to withdraw to Port Hudson rather than Vicksburg. He chose and fortified a position to command the Mississippi at Hudson. Superlative in strength, it remained in Confederate hands until Vicksburg fell. They gave it up only then because without Vicksburg it served no purpose, so well did Breckinridge do his work.

Union forces gave up Baton Rouge, for, adjacent to Port Hudson, it lost its value when the Confederates fortified Port Hudson. After the Union resumed the offensive up and down the Mississippi, the Confederacy also added and toughened fortifications at Natchez and Grand Gulf. Thus, because the Confederates had kept the crucial point, Vicksburg, they got back about 350 miles of the river, the stretch between the Yazoo and Port Hudson.

Events upon events in the West, taken together, demonstrated the valley's importance, and manifested the need to control the river, the river above all. That need and the river's significance had been lost on Henry W. Halleck. What if he lacked eyes to see enemy weakness? Did the defect matter when he lacked also the sense to make plans that would take advantage of the weakness? Besides, could he have made the plans, he lacked the energy to carry them out. Lack. Lack. Lack. What didn't he lack? He lacked enough to be a lackey, no more. Yet he still headed the Union armies as general-in-chief. Worse, he had not grasped what the river and the western theater meant. He ought to have supported Farragut at once and without question—circumstances screamed for it. Halleck ought to have committed twenty thousand troops, even twenty-five thousand. But Henry W. Halleck had to ride his hobbyhorse.

What did the Union pay for his blunder? As early as June 1862 the Mississippi and Vicksburg would have fallen and the Union accomplished with few losses what dictated bloody battles in July 1863. The hundreds of miles of river between the Yazoo and Port Hudson stayed Confederate because his blunders had left Vicksburg Confederate. So it took an extra year and its casualty-ridden struggle to secure the West to Union operations. Halleck also scattered an army ready for the offensive, 150,000 strong, on defense along a line stretching 150 miles. Then he became chief-of-staff in Washington. Grant, the new head of the Department of the West, found an army in sad shape.

For the time being he could not undertake anything significant. Washington through Halleck checked his freedom. He also lacked the strength, having sent a corps and other detachments east where Lee, advancing into Maryland, threatened Washington.

Confidence expanded and courage grew among the Confederates of course, what with such a state of affairs on the other side. Allowed a reprieve, they used it to strengthen Vicksburg. How they worked! They fortified the bluffs, reinforced Arkansas Point at the mouth of the Arkansas, and stiffened Fort Pillow as well as the other installations already named. General Braxton Bragg assumed command of the Army of Tennessee, and General John C. Pemberton the forces in and around Vicksburg. Bragg, having seen Halleck fritter Union forces, acted to gain by Halleck's mistake. Not Halleck but Bragg's new counterpart, Don Carlos Buell must pay for Halleck's mistake. Bragg struck Buell at Chattanooga and chased him to Louisville. Bragg installed himself at Chattanooga, the city on the Tennessee, a site fortified by Nature. Between Pemberton and Buell two Confederate independents operated: Generals Sterling Price at Fulton and Earl Van Dorn with men scattered between Baldwin and the Mississippi.

Bragg, maneuvering against Buell, brought them together. Don't let Buell's main force attack me, Bragg said to them. They, electing accordingly to attack Iuka and Corinth, captured both. General W. S. Rosecrans, heading the Union there, counterattacked. He wanted to dislodge them and did, in the Battle of Iuka, September 19, and the two days of slaughter at Corinth, October 3–4. Pemberton took over from Van Dorn because Van Dorn lost Corinth. The Union entrusted leadership to Rosecrans.

We have been speaking of 1862. The Confederates had occupied Corinth in March and April, and Bragg succeeded in Kentucky in August and September. But Confederate success peaked in the West then. Slowly at first, when the Union regained Corinth, but as surely as Grant massed at Grand Junction, the scales tipped to the Union. Only Sherman remained in Memphis in the old setup. Then, in the capital and the field, the Union shilly-shallied. In Washington the authorities wanted General John A. McClernand to capture the Mississippi on his own. Grant moved against Grenada in November and failed. He soon recognized Vicksburg's importance, however, and his headquarters drafted a plan.

The main force would go south on the Mississippi Central Railroad to Vicksburg. Meanwhile, at low water, troops with naval support would engage Vicksburg from the river. Grant assigned the river to Sherman. Washington assigned it to McClernand. Dispatches crossed. The mixup's consequences put Sherman at the head of the first contingent; his rival, the second.

By then (December) Grant's army, reinforced by recruits, had been trying to reach Vicksburg overland. His opponent, Pemberton, had set up behind the Yalobusha: Grant could not cross. Joe Johnston knew the weakness of Grant's supply line, the Memphis and Charleston, and the fragility of Grant's other lines of communication.[10] Johnston in consultation with his generals decided to cut them. Two units of cavalry set out, first the 2,500 to Columbus, Kentucky, under General Nathan Bedford Forrest. He left Columbia, Tennessee, on December 11. Rosecrans caught on and telegraphed far and wide. Watch out for Forrest, he said. By ferry, horses swimming alongside, Forrest crossed the Tennessee on the fourteenth and fifteenth (at Clifton), went to Jackson, then scattered the 700 Union cavalry at Lexington the eighteenth, capturing a fourth of them and the commander.

Grant at once sent generals to contain and capture him: Jeremiah C. Sullivan from Bolivar, William W. Lowe from Fort Henry, and Grenville M. Dodge from Corinth. Forrest (now withdrawing—feinting—eluding; now striking—skirmishing—fighting) routed two regiments of Sullivan's and hit one camp after another along the Mobile and Ohio, wrecking it and scattering its defenders. That man could lead cavalry! Two generals—Sullivan again (ordered out via Humboldt with 4,000 fresh troops) and Trenton, sent to Huntingdon—intended to intercept him on his return along the wrecked railroad via Union City to Dresden. He gave them the slip as best he could until December 31. They cornered him and made him fight then. He beat them but lost 300 men.[11] Still he trounced another regiment before returning over the Tennessee at Clifton, where he crossed the old way, by ferry, horses swimming alongside. He had to refloat the ferry first.

Van Dorn, like Forrest, had set out with cavalry to hit Grant. At first Van Dorn's 3,800 struck at Grant's lines with success to equal Forrest's. Grant with trouble and expense had set up at Holly Springs a base that, according to Union sources, cost a million and a half dollars. Van Dorn on December 20 attacked, captured the commandant Colonel R. C. Murphy and 1,500 men, and smashed what had cost so much. Van Dorn continued the expedition, but his luck had run out. Yet what he accomplished at the Springs cannot be overvalued. It said the worst to Grant: Forget trying to move an army to Vicksburg overland. See to your supply line; it counts above all. Withdraw north of the Tallahatchee, therefore, and be nearer the railroad. Repair it. Consider no moves south until you ready the railroad.

Sherman had been preparing an amphibious expedition. On December 12 he got the order on Vicksburg. He returned to Memphis and telegraphed Cairo and Admiral Porter ("Give me the navy's cooperation") and St. Louis ("I want transports to move 32,000 men downriver"). They complied so fast that a fleet of 76 transports, gunboats, and other armed craft, all steam-powered, left Memphis the twentieth. The 32,000 comprised 50 regiments and 10 batteries of 60 guns altogether, in 4 divisions under Morgan L. Smith, Andrew Jackson Smith, George W. Morgan, and Frederick Steele, all generals.[12] On the twenty-fifth they reached Milliken's Bend, above Vicksburg. They put A. J. Smith's on the right bank, to cut Vicksburg's supply lines. The divisions still aboard continued to the Yazoo and then up it as far as the heights of Walnut Hills and went ashore. Smith's rejoined them a day later.

Sherman, believing Grant could approach from the north, intended to land near Haines Bluff, surprise and take the hills, and go with him to Vicksburg. Should he not be there, Sherman would go alone. First, though, Sherman must secure the Yazoo.[13] A glance at a map will reveal the Yazoo, an obvious former Mississippi tributary, enjoying its own tributaries feeding its currents. Notice the upthrust: seventy meters steep from a shore of backwaters, inlets, and bayous, all prone to flood . Landing at Johnson's plantation, Sherman found only five, sometimes narrow trails leading among them to the heights.[14] Of course Confederate artillery commanded the trails. The men behind the guns could see every move, any gesture of attack. On December 27 he positioned his forces to fight. On the twenty-eighth they advanced into fire: A. J. Smith to the right; Steele, left; and Morgan and M. L. Smith (to be wounded), center. A bayou blocked

Steele. A narrow, brushwork levee led across but artillery raked it. He had to return to the Yazoo, take ship around Chickasaw Bayou, and land again, this time with Morgan's division in support. Confederates skirmished with them for the day, then retired to the heights. On the twenty-ninth, Sherman ordered them stormed: center leading, wings in support. Most regiments barely got out of the lowlands. Two brigades and a regiment, six thousand men, bulled to the Confederate line and attacked. They won high ground and held it—for a while. Under fire from every side, and finding themselves without support, they must yield, return to the low, and count for the day some two thousand casualties. Confederate losses did not reach two hundred.

Union men had just laid a plan. Those numbers, that outcome, prompted its reconsideration. Would Union forces ignore them, follow it, and within days attack the Bluff again? They decided right—in the pinch of Walnut Hills they could only be defeated—and they gave up on it. Sherman did not hold an enviable position anyway, facing the enemy from a swamp and having to bivouac where the river's rise could flood him out. He did not find inviting the high-water marks on the crowns of trees, four meters above a man's head. Camp here? He decided to bow to necessity and quit the expedition. Back to the boats he went, and down to the Yazoo to its mouth.

He arrived January 2, 1863, and met McClernand on the move from Memphis since December 28. McClernand showed him the orders naming him, McClernand, commander of the Mississippi expedition. Sherman put himself at McClernand's service. They decided to end operations here and having got Porter's pledge of cooperation, try to surprise Arkansas Point. They left Milliken's Bend the fifth and reached the Point the tenth. Porter's navy helped in the attack. On the eleventh the Point surrendered: field artillery, two Columbiads, and five thousand men. McClernand/Sherman occupied the Point. They wanted to divide their forces and continue. Grant said no. He ordered all under McClernand back to the Mississippi at Napoleon, Arkansas. Arriving there himself on the seventeenth, Grant called the expedition a wild-goose chase.

On January 26, Grant took personal, direct command of the theater. Guided by events there, he had decided on an amphibious campaign, intense, ambitious, and concerted. He intended to use the river to capture the river. He divided into corps—Sherman, McClernand, Hurlbut, McPherson—and ordered the first two, already in the

South, into positions opposite Vicksburg. Steam transports brought McPherson unit by unit to join them. Hurlbut occupied Memphis and held the railroad.

Farragut had tried but given up a canal through the peninsula below Vicksburg. Yet the obsession would not die; it reappeared in Grant's plan. An all-out dig began. Vicksburg's batteries zeroed in with an accuracy that knocked the idea cold. Doomed from the start, the dig continued all the same: to gratify public opinion as well as humor the government. On March 7 the Mississippi took pity, swelled, and ended the dig by flood.

Meanwhile the Federals had been seeking other ways to secure a landing above or below the city under siege. Two ways absorbed their energy and tested their resourcefulness. (1) Could they open a passage to Lake Providence, thus leave the Mississippi via Macon Bayou for the Tensas and Red Rivers, and thereby return to the Mississippi below Vicksburg? (2) Could they cut the levee that separated and blinded Yazoo Pass from the Mississippi, and go down the Yazoo to Vicksburg? They, those Federals intent on Vicksburg, worked at both with vigor.

We shall return to these engrossing events. First a glance at the defense. Joe Johnston, though east near Chattanooga, monitored the Mississippi: each act, every detail. Pemberton, headquartered at Jackson, commanded the department.[15] General Carter L. Stevenson headed 23,000 at Vicksburg. Another 15,600 under General William W. Loring held Granada and guarded the Yazoo. At Jackson, 4,000 had to spread themselves among several positions besides duties such as protecting the railroad. Van Dorn and his 6,000 cavalry, sent to Chattanooga, followed orders to establish a presence along the Tennessee and then join a thrust north of Corinth.

The Federals intended to pierce those defenses. The map shows how the newly conceived Lake Providence waterway would carry troops by a long route: about 250 miles to the mouth of the Red, and another 200 up the Mississippi to Vicksburg. Though a mere 7 miles divided Bidal Bayou from the lake, the dig meant toil and torture for the engineers. Stumps, filling a swamp, must be grubbed and disposed of; then the levee between the Mississippi and the lake must be cut. McPherson, in charge, could not bring in vehicles, tools, and equipment until nearly March. The digging and grubbing proved harder than expected. Reaching the levee at last, the diggers and grubbers found the digging and grubbing easier. Water poured through on March 18. Boats, however, must wait; shallows and logs blocked navigation. By the end of March, McPherson reported to Grant a canal open to boats no more than two meters in draft. Underwater sawing apparatus, the only means possible, had dealt with the 12 to 15 logs while dredges gnawed at the shallows. Meanwhile the high command had given up on this project. Yes, the achievement I have just described, the Lake Providence way, done, would not be used. The Union abandoned it in favor of Yazoo Pass.

The Yazoo, its upper part called the Coldwater, bends and twists toward the Mississippi and Helena, until it has approached within thirteen miles. Yazoo Pass—we have called it a bayou—continues from there. Like the route just finished, it could take boats of shallow draft. Indeed, earlier, it had been the waterway between Yazoo City and Memphis. Later a levee cut the Pass off from the Mississippi, to improve Mississippi navigation and control floods. The levee had raised the Mississippi above the lowlands around the Pass. Therefore only pierce the levee, reinstate the route, and open Vicksburg to attack from the north. Thus the theory behind the Yazoo Pass project.

During work at Lake Providence, work at the Pass had been proceeding and with more energy and greater devotion. The levee—six meters high and thirty wide—must be cut. On February 3, explosives sliced through. The river poured into the lowlands. By the seventh the torrent subsided to admit the first gunboat. It floated but could not penetrate the tangle of tree limbs. Confederates, finding limbs overhanging and using the time well, had bent some into the water and lopped others to thicken and complicate the obstruction. Federals had to labor until February 21 to let boats into the Coldwater.

Grant, hearing of the progress, ordered General Leonard F. Ross's division to Helena, to destroy from there the railroad at Grenada and reconnoiter the route through the Pass. The order went out February 15; the division did not board transports and move until the twenty-third. Porter's escort accompanied them: four wooden yawls with cannon, two ironclads, and a ram. The expedition cleared obstacles and survived adversities, mishaps, and casualties. What with every boat damaged, however, and some almost crippled, they did not arrive on the Coldwater until March 2. Yet reports to Grant praised the route. So he decided to increase McPherson to 30,000 and send him down the Yazoo to Yazoo City and a comprehensive landing there. Grant also ordered two cavalry units to cooperate, one from Helena east, the other from La Grange south, to meet at Grenada.

The Confederates, informed by an excellent intelligence network, had not been idle while the enemy prepared. The party sent to lop limbs and block the Pass—remember them? They assembled floating batteries on the Yazoo besides. Pemberton, learning that the enemy would send boats through, directed Loring and plenty of troops to the mouth of the Yalobusha to prevent passage beyond it. Loring reached Greenwood on February 21, promptly enough. The Tallahatchie concerned him, too, as it emptied along with the Yalobusha into the Yazoo there. He sank hulks in the Tallahatchie's mouth. To keep in touch with Grenada he left the Yazoo and the Yalobusha open. Heavy artillery arrived, eight cannon sent up the Yazoo from Vicksburg. Opposite Greenwood he erected the so-called Fort Pemberton. This small but well-sited works commanded the upper Yazoo with the eight cannon from Vicksburg and two thousand men behind cotton bales and earth.

The fort had been completed on March 10—just in time. Ross's division, in boats, appeared on the eleventh. Ross had intended surprise. Obstacles, raised by the enemy along the Yazoo and Coldwater, delayed him from start to finish. He surprised no one. Worse, he had brought four thousand men, more than enough, but could not land them. Conditions underwater and on the bottom kept them away; the boats must mount the attack. The first day saw a few shots exchanged. By the twelfth of March he had got a battery ashore, six hundred meters from the fort. The few shots became a bombardment. Each side pounded the other, the boats suffered, but the works showed not a scratch. No choice for him but to quit. He steamed back up the Yazoo the way he had come. On the twenty-first, he had gone far enough to meet Isaac F. Quimby's brigade. It had finally got into a few boats and underway on the new route after a delay in Helena's environs. Quimby, the senior general, ordered Ross to join him and turn back. Together they would try again to take Pemberton.

They arrived on the twenty-third, to try their luck once more. The guns afloat failed against the guns ashore. Riverfront conditions for the second time thwarted the troops' attempt to land. Union luck had not changed: none ten days before, none now. Quimby, old plans foiled, made new. He would throw a pontoon bridge across the Yalobusha, go east, and flank the fort, only he had no pontoons.

He sent for some on the twenty-seventh. On April 1, the steamer with his order met a steamer carrying an order for him. Quit, it said, and return to the Mississippi with every man as fast as possible.

Truth had dawned on the people who gave him the order. It would require masses of troops to take Pemberton. And, worse, conditions called for small boats to land them, boats that could not be found. In addition, Grant had been getting reports of Confederate reinforcements to Grenada, and of the Confederates adding equal starch along the Yazoo and at Greenwood. It amounted to such a threat that he could not secure the supply lines in the Yazoo region. Acting therefore on fears for Quimby, Grant decided to recall him before disaster dispatched him. They must give up Yazoo Pass and the notion of Vicksburg by one amphibious thrust from the north. Grant chose the so-called Steele's Bayou project instead.

To understand it, go up the Yazoo; take any of several passes to Steele's Bayou; follow it thirty miles to Black Bayou, and Black Bayou to Deer Creek, Rolling Fork, and the Sunflower River. In 240 miles the route has brought you back into the Yazoo, 60 sixty miles downriver between Yazoo City and Haines Bluff. You have gone Grant's route when he joined the reconnaissance that learned how a single saw to a few trees could open or shut the way for any and every steamboat. You observe, too, that in the cross-channel narrows, I mean in Black Bayou and at Rolling Fork, boats could not turn around. A saw in the hands of the enemy and put to a few trees could block the way for many if not every steamboat. The Federals might cut their way out but, after the cutting out, could they hope to reach the Sunflower and its generous channel undetected?

Sherman and part of his corps (three regiments) and Porter's escort (five ironclads, four mortar boats, and two tugs) followed the plan and that route. Driftwood in heaps, and further hindrances, and blockage upon obstruction, had to be removed. Another regiment, recruited mostly among Mississippi boatmen, went ahead therefore. So they proceeded—the advance guard clearing the way—through Black Bayou, undetected and happily so. Sherman fell behind. Porter neared the Fork, advance guard at work. From all sides and all of a sudden, artillery rang out. Porter sent every available man ashore to fan out and block whatever might be the enemy's intent. This spontaneous defense, too weak to stay and survive, near evening must return. The situation grew critical. By morning the bushes and trees swarmed with Confederates. They drew a bead on any Federal who dared show himself on the boats. To the front and rear the Confederates felled trees and flung them across to block the strait. They sank a coal barge: another block. So they worked unabashed, doing everything to bottle up Porter's boats and take him and his men prisoner. He, the night the trouble started, sent a Negro with the word to Sherman: danger. Porter decided he must wait for Sherman but destroy the boats and—to save the men—lead them through the thickets to the Mississippi. Sherman had eight hundred men with him. Them he sent forward at once while he hurried to the rear to hustle the rest after them. By boat they steamed full ahead as far as Black Bayou, debarked, and followed guides by torchlight through dense canebrakes. Porter had begun the retreat when Sherman's first troops arrived in his behalf. Sherman meanwhile discovered the enemy about to seal off the boats for good. He dispelled the threat. By morning he reached the embattled flotilla and liberated it. Away the boats went, only having to keep pointed upstream while backing downstream. The Steele's Bayou project could no more succeed than they turn around. On March 27 they regained the Mississippi at last, aborting the doomed brainchild.

Three projects—Lake Providence, Yazoo Pass, Steele's Bayou—had failed. The Union army and especially Grant's headquarters as well as the people of the North had pinned hope to them. Everyone's hopes had been dashed. Public discontent flared. Yazoo Pass in particular riled the populace; so much had gone into it.

Grant did not flag. Reconnaissance had discovered that the Walnut, Roundaway, and Bidal bayous led to the bayou at Milliken's Bend and therefore, as a map shows, could link Duckport and Hard Times. You have, you see, only to dig a canal (so the reconnaissance said) between the Mississippi and the bayous. Merely one canal, between Young's Point and Duckport, would establish communication between Hard Times and Duckport. A campaign would go from Hard Times across the Mississippi, scale the bank, and take Grand Gulf, approaching from the south then, to attack Vicksburg. The plan called for thirty thousand troops to assemble at New Carthage while Steele's division feinted along the Yazoo to divert attention. Water ran through the canal on April 13. Dredging until April 22 deepened it for small steamboats. Then water in the Mississippi began to drop. Less there: less in the canal. Water diminished until nothing but rafts—rafts drawing the minimum—could find enough. One more project done for!

The Milliken's Bend canal could float nothing but rafts yet led to the heart of Vicksburg. Grant, impatient with delay at the dig, had sent troops of McClernand's corps overland across the Peninsula to New Carthage. Despite difficulties, they got there. Everything the attackers would need, field artillery too, could follow: so much the corps had proved, being the first to take the route. As for rafts on the canal, they proved that if good for nothing else, the canal would carry supplies on flatboats. Every strength and the utmost energy went into dams, bridges, and other improvements to make the ways passable and the attack possible. As early as April 23, many regiments had reached New Carthage. Grant arrived. Finding it surrounded by water, he decided against it as a place to assemble for an attack. The cliffs opposite, manned and fortified by Confederates, ruled out a Mississippi crossing anyhow. He massed at Hard Times. Every condition favored attack from there. Even the crossing posed no danger, being far enough below the enemy's fortifications.

The boats must be brought up now to guard the crossing and serve as the supply base. Of course, should the worst happen, and the army need an out, the navy could ensure the retreat. Boats of the line, drawing too much for the bayous, must imitate the *Carondelet* at Island Number 10. I mean they must pass Vicksburg's batteries by night: in the American phrase, "run the gantlet."

They ran it the night of April 16/17. The way had been tested, first by Farragut in 1862. On February 2, 1863, Colonel Charles R. Ellet's *Queen of the West* passed in the morning in broad daylight. The *Queen* had not expected the delay that had made her late; she intended to go the night before. Worse, against Vicksburg's artillery she must accept cotton bales, not assert armor. Vicksburg fired nonstop for fifty minutes. The *Queen* took twelve hits and lost a gun. The barrage also smashed the cabin. Still she remained able and ready to fight. Finding the *Vicksburg* at anchor under the guns of the city of her name, the *Queen* tried to ram and sink her in passing, with partial success.[16] Safely past, the *Queen* accompanied the Union *DeSoto* downriver. Later she captured several Confederate supply boats on the Red. On February 13 the Union gunboat *Indianola*, Lieutenant George Brown, passed Vicksburg, imitating the *Queen*. On the sixteenth the *Indianola* met a small steamboat with Ellet and his crew aboard. Ellet had lost the *Queen* in a duel with a battery along the Red.

After a pause around the Red, the *Indianola* dawdled past Grand Gulf and back toward Vicksburg. Meanwhile on the Red the Confederates raised and repaired the *Queen*. They also fitted out the ram *Webb* and gave her two smaller cotton-bale–padded companions. Readied thus for the *Indianola,* the four hunted her down. On April 24 they attacked in the dark when her excellent artillery could not be sure of a target. Inferior to her in every feature except that they could ram, they took advantage of her night blindness, rammed, and nearly sank her. In a rush her crew pushed the guns overboard and drove her aground. They leaped ashore—to be captured. The Confederates controlled the river between Vicksburg and Port Hudson.

They kept it only until March 14, when Farragut with the famous *Hartford* and the smaller *Albatross* ran the batteries at Port Hudson. On March 19 the pair had gained the river below Vicksburg. In a vulnerable position, Farragut considered the pair too weak to hold it. Send an ironclad, he urged Porter. He wanted to stop the work at Warrenton, too, he added, because the Rebels gained strength there every day, expanding the batteries. Porter sent the *Lancaster* and the *Switzerland*. Delayed, they did not reach Vicksburg's guns until dawn, March 25. The unhappy *Lancaster* took enough hits in passing and blew up. The pounding disabled the *Switzerland*. Those misfortunes left Farragut little but to take the pair back to the Red.

His efforts and the others', however, taught important lessons. Armored gunboats needed nothing but steam to run shore batteries. Boats without armor could run them, too, behind cotton bales. Preparations must be made and precautions taken in time; a run would do best under cover of dark from beginning to end.

Lessons learned, the squadron prepared. Ready on April 16, they weighed anchor at the mouth of the Yazoo, 10:00 P.M., and moseyed down the Mississippi. Porter led in the *Benton,* the command boat, sixteen guns. Next the *Lafayette,* Captain Henry Walke, ten; and lashed to starboard, the captured *Price.* Then the *Louisville,* Captain Elias K. Owen, thirteen guns, followed by the *Mound City,* Captain Byron Wilson, thirteen; then the *Pittsburgh,* Lieutenant William R. Hoel, thirteen; and the *Carondelet,* Lieutenant J. McCleod Murphy, thirteen. After them the steam transports *Forest Queen, Silver Wave,* and *Henry Clay,* full of food and munitions, and padded to depth with bales. The gunboat *Tuscumbia,* Captain James W. Shirk, brought up the rear with three guns. Each towed a barge of coal, often brimming with 1,000 bushels. A point jutted into the river at Vicksburg. By 11:10

the defenders spotted the *Benton* passing it. They started shooting. At the same time, on signal, along either shore, they put the torch to house after house and heap upon heap of combustibles. The squadron began the perilous transit as if in daylight. Each boat must endure shot, shell, and bomb—a fury of burned powder—in a gantlet lasting about a half hour. The boats in self-defense hugged the shore and poured broadsides into the streets. At 2:00 A.M. the squadron reached Hard Times and anchored, for the most part undamaged, only a few barges lost. The *Henry Clay* had gone up in flames, but the crew got ashore. Casualties amounted to ten or twelve, wounded but slightly, nothing more.

What had the Confederates been doing before that Union success? They had dismissed Union naval prowess, if it had even impressed them. They had come to such a persuasion because they believed the Union navy withdrew after the Yazoo failure, to Memphis or even farther north to lick its wounds. Vicksburg therefore had not been readied enough against attack. Two commanders must be held responsible: at Vicksburg itself, Pemberton; on the right bank, across from the city, General John S. Bowen, supposed to watch for Grant and harass his advance. The two had convinced themselves that what happened on the peninsula between Milliken's Bend and Hard Times had not been important: feints, they thought, by scattered detachments. Pemberton, blinded by the misconception, reported it as fact to Johnston. Units accordingly left for positions on the Tennessee, by orders from higher up.

At the end of March, Pemberton still had about 50,000: Carter L. Stevenson's corps, 22,000, in and around Vicksburg from Grand Gulf to Haines Bluff; in addition to Franklin Gardner's 16,000 in Port Hudson; plus William W. Loring's 7,000 at Grenada, Fort Pemberton, and elsewhere; and Bowen's 2,500 and 4 batteries.[17] Confederate lines reached from Fort Pemberton to Port Hudson, 125 miles. Fortifications had also been extended south to Grand Gulf. Bowen across the river made comfortable reports about the peninsula, as I have said. They prompted Pemberton to send at Johnston's request 8,000 men east: 1 brigade in early April, another on the fourteenth. The second had just left when the fleet passed, clarifying the situation soon enough to call them back. The first had been gone too long to be recalled. Even Bowen on the Louisiana shore appreciated the operation's size, albeit not in time. He brought his detachment across with difficulty but unscathed.

Grant's operations had grown darker and darker: more obscure, more secretive. Lacking cavalry, Pemberton would never find them out now. Confederate confusion increased when a bold, effective raid brought Colonel Benjamin H. Grierson's Union cavalry to the gates of Jackson and left the defenders bewildered. Pemberton's response with infantry, as useful against cavalry as a baton against mosquitoes, managed only to heap perplexity upon confusion. And then the report to Pemberton: Troops in number have landed at Hard Times. In response he recalled a brigade from Grenada to Jackson, and posted Bowen with 4,700 south of Vicksburg. Still the Confederates suffered from a jaundice that Union action below Vicksburg meant little, feints, they thought, ruses to distract us from a thrust along the Yazoo.

So we see a brigade beyond recall (a blunder) and the cavalry sent elsewhere (worse) and a Vicksburg without its eyes and ears, as it were, spelling disaster for the Fortress of the West and sealing the fate of the South.

Grant meanwhile, overland and by steamboat, had massed and concentrated near Hard Times. He intended to cross the Mississippi at Grand Gulf and encircle Vicksburg from the south. Grand Gulf, a fortified village, commanded the river from the mouth of another, the Black, navigable a few miles up. On the Black's left, nearly to the Mississippi, bluffs rose as high as forty meters. Two batteries, thirteen guns in all, anchored Grand Gulf's armament. One aimed upstream, one down, they covered the site and must be silenced else he could not land. It seemed as if he could silence them with a squadron of eighty-one guns, including eleven-inch smoothbore Columbiads and nine-inch rifled Dahlgrens (corresponding in our terms to artillery of twenty-four- and thirty-centimeters). He could measure up to the thirty-two-pounders on the other side. The boats sailed in a group—Porter's gunboats first, then the transports with three McClernand divisions, and a squadron from Hard Times—April 29, 7:00 A.M., to force the crossing. The gunboats squared off against the Confederate works. The transports, out of range, unloaded on the Mississippi's shore. Grant commanded from a small steamboat in the middle of the river.

I attach a letter by Porter, to illustrate the disposition of forces in this amphibious attack.[18] The bombardment, on the plan to a T, began at 8:00 and ended at 1:00. Both sides let fly with a will. Into the fresh earthworks Porter hurled twenty-five hundred rounds that burrowed into them and plowed them up, and harrowed, ripped, tore, and gashed them, and seemed as if they must have silenced every last gun and wiped out all signs of life. In truth a mere three Confederates lay dead and only thirteen wounded. Among them, however, the chief of artillery, Colonel Wade: killed. Porter's boats suffered several hundred hits, nineteen dead, fifty-six wounded. The *Tuscumbia* suffered most: eighty-one hits. A few pierced her armor and exploded in the hold. She had to give up and withdraw. Even Porter's command boat, the *Benton,* took forty-seven hits and lost twenty-six men.

The shore, outgunned, had shut up but remained unbeaten. The artillery could fire; the earthworks, for all the uprooting, still served the purpose; and the garrison would defend itself. The afternoon proved it stronger than the attackers. The transports waited opposite, crammed with troops that dared not go over and land to attack. Grant gave up at Grand Gulf.

If the packed transports could not land troops at the still-dangerous Grand Gulf, even less could they pass unhurt. Grant ordered the three divisions off where they got on, at Hard Times, to march to DeShroons. The transports, empty, must pass those menacing guns again. Therefore he sent the transports in the evening, supported by renewed naval bombardment. Neither the darkness nor the artillery covered them; they passed but got hurt. In the morning of April 30, everyone had rendezvoused, after all. Lacking cavalry, the Confederates could not well resist the rendezvous, let alone prevent it. Grant, no enemy to oppose him on the shore opposite, had only to send McClernand's forces across, every last man, including the Fourth Division just arrived. They landed without mishap. As soon as the evening of May 1, they had reached Port Gibson. From there, Vicksburg could be stormed.

Further events in and around Vicksburg—siege, storm, capture— serve our study only in general. Confederate ranks had thinned to 17,000. Grant pressed. Grand Gulf yielded on May 3; Sherman joined Grant on May 7; the Confederates kept retreating; and in Mississippi, one after another Confederate post fell in skirmishes and battles: Raymond (May 12), Jackson (14), Champion's Hill (16), and Black River Bridge (17). When Haines Bluff to the south gave up (18), the ring closed. Grant made overtures—Pemberton refused. Grant attacked—Pemberton beat back the storm. Grant must stand off, forced to do siege, partly because the brave Pemberton inflicted terrible losses. But, lacking resources, Pemberton could not hold out. Johnston tried to relieve; every try failed. On July 4, Pemberton had to surrender every man and the city itself. With Vicksburg the Confederacy lost the keystone of resistance in the West.

## 5

We have discussed Grant's struggle to overpower Vicksburg. Meanwhile the Confederacy tightened its grip elsewhere on the river, and the Union determined to break it. In 1862 the Union captured and fortified Baton Rouge. That strategic site controlled the Mississippi of course, but also the Red at the mouth and a stretch upstream. Therefore the July 1862 order to General John C. Breckinridge: take back Baton Rouge—surprise attack. This clever, able Rebel knew warfare—he would be the Confederacy's last secretary of war—but Union forces beat off his attack, spilling blood by the bucket. He chose a practical alternative: fortify somewhere else. Port Hudson would serve instead and comply with his orders in spirit. Indeed, lying closer to the mouth of the Red, Hudson might be better than Baton Rouge. By the start of 1863 the Confederacy had used the six months well, arming Hudson to the teeth. We have reason to believe the numbers: a garrison of eighteen thousand.

In November 1862, General Banks left New York with reinforcements for New Orleans, to relieve General Butler in command there, and to cleanse the lower river of Confederates. Banks by mid-December had sent ten thousand (General Cuvier Grover's brigade) to Baton Rouge. They routed a Confederate outpost nearby, but Grover did not think them strong enough for Port Hudson. Banks as well, with but fifteen thousand, dared not risk a front assault. He knew that Hudson's food and other supplies came east on the Red; he had better interdict those first. He tried from south of Grand Lake: the expeditions never got beyond the line Opelousas/Plaquemine. That terrain—every feature, every inch—favored the enemy, at home there. Indeed, Banks met an enemy who controlled the right bank of the Mississippi from Alexandria at the Red, down to Grand Lake, and to the Gulf, with fortifications in the swampy reaches: Fort De Russey just south of Alexandria, Butte à la Rose at the Atchafalaya, and Bisland on the south shore of Grand Lake, all under General Richard Taylor.

Banks had been stopped. Union gunboats *DeSoto* and *Queen of the West* had been destroyed.[19] The loss alerted Farragut: Rebel control of Port Hudson and the Red meant danger. Farragut suggested that he, Banks, gesture at Hudson while he, Farragut, tried to take the fleet past. Misunderstandings spoiled coordination. Farragut without shoreside support ran the batteries the night of March 14/15, 1863. We must look closer; the episode merits study for what we want.

The fleet went in this order, mostly in pairs.

1. *Hartford* (command boat), sloop — 28 guns
   *Albatross*, gunboat — 8 guns
2. *Richmond*, sloop — 25
   *Genesee*, gunboat — 8
3. *Monongahela*, sloop — 11
   *Kineo*, gunboat — 6
4. *Mississippi*, sloop — 19
   *Sachem*, gunboat — 5
5. *Essex*, ironclad — 7
6. Five mortarboats (one gun each) — 5

If those yoked fast in pairs could better negotiate the current, little else favored passage. Mishap, accident, disaster befell nearly every boat. Enemy fire pounded *Hartford/Albatross*. The pair arrived upriver of Hudson—after spinning in a circle, nearly ramming *Richmond/Genesee* at the same time, and running aground. The *Richmond*, hit in a steam vent, lost power and had to drift back downstream, dragging the *Genesee*. They anchored below Hudson. *Monongahela/Kineo* struck shore, ran aground and, damaged by gunfire, must retreat after *Richmond/Genesee*. The *Mississippi* ran aground too, in shallows. Captain and crew, unable to get her off, destroyed everything useful, took the valuables, and left. Empty, she drifted, caught fire, and exploded. The *Essex* gave aid and comfort to the damaged. She therefore, as well as the mortarboats, never passed Hudson. The expedition failed, by and large, Yet the fraction of success accomplished the mission. *Hartford/Albatross* passed Hudson, reached the Red, and interdicted traffic as intended.

In the spring of 1863, Banks had little to fear for New Orleans. Grant could expect and would have been happy with his cooperation against Hudson. But messages, often sent along the west shore of the Mississippi, met on that parlous overland route too many troubles and excess adversity on those risky paths. Information, reports, suggestions, and other communiques—practically every dispatch crossed one it ought to have anticipated. Neither understanding nor harmony nor cooperation from these communications, therefore, but misunderstanding upon friction upon discord. Besides, each general took orders from Washington, and acted, not in accord with one another, but according to Washington. At last in May of 1863, orders came from higher up, to Banks, for an assault on Hudson.

General Franklin Gardner commanded there. By late May the threats to Vicksburg prompted him to send every man he could spare to its aid, leaving him with but 5,000. Banks attacked them with 33,000. On May 25, Banks and that horde invested and tried to storm Hudson. Bravely, Gardner and his defenders repelled the attack and took toll of the attackers. Banks on June 13, though he had lost 2,000, called upon Gardner.

"Surrender."

"No."

Banks tried again to shoot his way in, lost another 1,800, and managed only to advance his siege works a bit. The gain, however slight, put his forces around the little fort. Tightening the noose, they had squeezed close to the embankments when Vicksburg fell. Grant's troops could be added to those choking Hudson. On July 9 the puny garrison surrendered; the last Confederate presence on the Mississippi ended. The Union had freed the river for the Union's use, exclusive and unimpeded. The Confederacy tried to gain another hold once, in the assault on Helena from the west: beaten back, bloody. The Mississippi belonged to the Union and nobody but the Union.

## 6

Confederate forces had to surrender Fort Henry, but neither the enemy's army nor his navy deserves the credit. The brave defenders surrendered because their leaders blundered. The Confederate army's high command missed the region's worth; thus the valley's significance and the river's importance remained unknown, alien, to them. Instead the Confederates exhausted themselves in operations of secondary consequence. In the East, as I suggested at the outset, Confederate leaders planned and directed campaigns with savvy, and carried them out with vigor and panache. We seek but do not find the same in the West. Joe Johnston may be the exception, but his leadership could not offset other's mistakes.

At Donelson the navy took no direct part but aided victory by bringing up provisions, moving troops, and the like. Panic, uncalled-for and inexplicable, must have caused surrender. Why else such disgrace by forces adequate to the defense? The Confederacy suffered in proportion to what it lost: a position of significance, priceless. The loss disappointed southerners. Discontent spread over the Confederacy.

Commanding generals changed in the West then: the North's to Henry W. Halleck, a painstaking but cautious, hesitant, even squeamish leader; the South's to P. G. T. Beauregard, known for victory at Bull Run. Beauregard, sizing things up after the losses of Henry and Donelson, rightly ordered withdrawal from Columbus, Kentucky; and the Confederates left on February 26, moving their line south. From Memphis it ran via Shiloh to Chattanooga now. Two fortified points remained as outposts on the Mississippi: New Madrid and Island Number 10.

Confederate forces in the East lived and fought by discipline. What a jolt to see it and then learn what happened in the West—the disgrace, the ignominy, the shame of officers at Fort Donelson and men at New Madrid. When low morale, depression, gripped men one and all at New Madrid, officers ought to have acted. Officers and men alike—with strength, by effort, with devotion—ought to have defended the place. Yet think of the officers' duty, how hard it had become. What to do in an army of green militia that gave officers little authority? Think, too, of service in a place like New Madrid. Could duty seem urgent in an assignment so unhappy?

Measures ought at once to have followed the loss of New Madrid, to secure Island Number 10 against a similar surprise from downriver. The island's defenders fought well enough, but the Confederate West lacked reliable, perceptive, decisive leadership. The measures such leadership would have taken therefore did not follow the loss of New Madrid. Island Number 10 fell; lame leadership did not comprehend the significance of the theater nor grasp the importance of being able to interdict the river.

How different the Union in the West! How we should heed its vigor, devotion, and courage! The travail and the agony of canal digging, and the joy the army asserted in the face of such hardship, could we forget those marvels? Morale must be high, devotion to duty profound, when engineering and the rank and file try to outdo each other: the one to devise machines to remove obstacles, and to dig and excavate; the other to persist and persevere to finish a labor almost superhuman. And then the navy equaled that zeal of patriots, Captain Walke above all. He dared first the shooting gallery where cannon ashore drew beads on ships passing like so many ducks in a row. Imitating him and his *Carondelet,* the fleet ran that array of fire-spitting ordnance. His intrepid act helped the army advance in the West—advance and win.

Let us not praise one and another branch of the forces and miss Island Number 10's lesson for military science. It restates the principle, a truism by now, that the navy cannot achieve large-scale objectives even against weak fortifications. Success demands army-navy cooperation: combined operations.

The side that held the Mississippi would hold the strategic advantage. Confederate commanders, I have said, grasped the fact too late. It followed that they did not do enough for Vicksburg, especially early in the struggle for the river. Nor had Confederate troops anywhere west shown the energy and heroism of counterparts east. (I make these statements after a study of Confederate documents.) Hence a Vicksburg unwilling and unable to defend itself. One threat to the practically open city would probably have put everyone to flight.

All the more reason to condemn Farragut. True, the ever-enterprising admiral deserves laurel upon accolade for capturing New Orleans. But he came to Vicksburg the first time without the same confidence, pessimistic. He did not try the *coup de main* that could have saved 100,000 lives. *That* many? The statistic of those who died in the ensuing campaign for Vicksburg, no exaggeration, represents only Union men, struck down by saber, powder, and ball, and by sickness.

Things changed before Farragut ran Vicksburg the second time. The first, a reconnaissance, had alerted Vicksburg's commanders. In the interval, trying to turn a town into a bastion, they exerted every effort to increase the garrison, strengthen the works, and reinforce the armament to the extent Confederate resources could supply the big guns. Meanwhile, Union leadership failed again, this time in the high ranks. Farragut used what he had; an assault on Vicksburg deserved—demanded—more. Events would betray the seriousness of their mistake. My readers know how, despite Confederate leaders' efforts, Vicksburg's works remained bad and the artillery worse. Yet they proved obstinate and effective against the excellent artillery of the fleet.

Indeed, striking back, the Confederacy took the upper hand for a while. Events confirmed the importance of the valley while proving that the navy must cooperate if the army shall stand its ground and make war in the West. My readers see the Federals increase to superior numbers yet stay inferior on land. The numbers favor Halleck 3:2, Grant tries campaign after overland campaign, yet neither landbound chief prevails. Let the Federals touch water, however, and they seem transformed; they gain power; they become Titans. They learn: the fleet must help. Combined operations will overtake the Confederates and bring the Union victory.

Halleck lacked a general's virtues. I have described him as neither born with the energy to force a good result out of a bad plan nor granted the eyes to see an enemy's weakness. Initiative had not been thrust upon him, either. He wanted the energy, the determination, the enterprise to take obstacles in stride. He demonstrated noetic poverty early, slouching toward Vicksburg to no effect but the strengthening of the Confederates. Worse, to the north, where the Union held the all-important Memphis and Charleston, he did not smite the Confederates with his superior number and keep the foe from the rails. He frittered his forces along its length—Sherman at Memphis, McClernand at Bolivar, Pope at Rienzi, and Buell east of Iuka, put there and commanded by him from Corinth—all trying to repair what could not be repaired or would not stay repaired. Scattered and fragmented, his forces could not stop the raids that kept causing the damage.

He could be grateful that the Confederates—Pemberton at Vicksburg, Bragg in Tennessee, Price at Fulton and Iuka, and Van Dorn between Baldwin and the Mississippi and at Corinth—suffered the same disorder. Wanting to be everywhere, scattering instead of centering, not rallying at a point and engaging the enemy from there together, trying to hold everything, they held nothing. True, like a malaise, awareness had crept upon the government at Richmond that unity suffered in the West. President Davis himself, to promote unity there, traveled to the Mississippi with Joe Johnston. Yet to what avail did Davis put him in charge between the Alleghenies and the Mississippi? Johnston's lieutenants had independent assignments, therefore autonomous commands, making the question vexing. He told Davis again and again that he, Johnston, held an ambiguous office. Events would bear him out.

When Grant's forces hit Vicksburg the third time, they failed for too few troops, inadequate cavalry, and defective logistics. Then came Sherman's harebrained Walnut Hills expedition, innocent of reconnaissance and short on thought. In other words, Sherman's dare lacked every requisite for success. The Confederates had elaborated a net of intelligence and communications. Its excellence precluded surprise. Nor could Sherman expect cooperation by Grant and the main force in movements from the north (another evidence why we call the scheme harebrained). It remains a mystery, too, Sherman's expecting opposition from 6,000. Yet he might as well have opposed such a force. Ahead of him, north of Vicksburg, 1,200 had concentrated, dug in on the heights, and made ready to fight an enemy five times their number. Up there they had the Mississippi and its valley at their feet. Able to see every movement in it, they could decide in plenty of time what to do: set a trap, say, or prepare a red-hot welcome.

The plan that in outline called for simultaneous attack by land and water—the soundest so far—failed for enough strength to leave a corps to the north to hold Bragg at Chattanooga. The fourth strike, ending therefore in defeat, turned things upside down. It seemed to wreck plans for an overland campaign. Planners and leaders began to reconsider what they had done, and to rethink what they would do. Then the tragedy of Holly Springs produced the desire for a line of supply that could not be broken. Planners and leaders saw that the Mississippi would provide it. Success at Vicksburg demanded attack by a safe, fast route, too. The Mississippi would be it. The best operations must be independent of weather, moreover, and not at the mercy of roads without bottoms. The Mississippi would be the savior. The fleet offered a further advantage. Aboard its boats the infantry and the artillery, the big guns as well as the food, needles, and thread, and the powder, shot, and shell, could be assembled at the attacker's convenience, brought together at his beck and call, and used when and where he saw fit.

The furor and bustle of the Lake Providence, Yazoo Pass, and Milliken's Bend projects, and the Federal desire to control the valley, besides the Federals' peculiar expression of fervor and fire to get what they wanted—need I mention those factors again? By now the Federals understood that naval support produced a geometric increase in army strength. In this period I dare say they overestimated how well the navy could fight. Yet they missed the arrangement that could do the trick: combined operations. Grant would have been smarter to repeat what had been tried in the Walnut Hills episode. I mean he ought to have sent the army on foot from the north and moved the navy on the river: a combined operation to strike Vicksburg by water while storming it by land.

Not defective plans, however, but simpler and more basic faults contributed to early failures. *Item:* Grant and Sherman, communicating neither by telegraph nor by less direct methods, lost touch. *Item:* Lincoln's cabinet, growing more and more bizarre in deliberations, gave two men the same command, Sherman and McClernand, both to head Mississippi operations. Still, failure to take Vicksburg from the bayous lay less in ill-laid and badly hatched plans than in the death that bristled from the bluffs. No Union plan, nothing in Union power, could cope with the Confederate weapons on those heights. Review what happened if you want to sum it up. Let the events tell the tale.

Grant, as an American author observes, had come to the moment of truth.[20] Not yet known, let alone renowned, he stood among Hooker, Rosecrans, and others of the second rank. Victory at Donelson and the draw at Shiloh brought recognition. Then Memphis, Iuka, and Corinth: twelve months of fighting in vain. Then three months along the river: sweat and strain, mishaps and failures, untold lives lost, and not a tincture of success. Public opinion derided his plans. "Crazy," people said, "worthless." Doubts in the army, too, about "the right man for the job." Gloom gathered around him as general. Blight settled on his career. Shadows darkened his name. Yet, deserving of credit and a good word during that time, he accepted failure and did not complain. Devotedly, doggedly—with fidelity, with tenacity, with passion—he carried plan after plan to its end. Setting an objective once and for all—pursuing it—worrying at it—

persisting—helped win at Vicksburg and later on the battlefields of the East.

He took Vicksburg at last because he put the troops ashore. He could have got them into position easier had he landed them sooner. We recall what happened in the East. Richmond could be taken only via the peninsula and because army and navy could cooperate there. McClellan recognized the fact as early as 1862. Grant in that way finally took Richmond three years later. Indeed, the two campaigns in this sense differed in only one respect. The Union could secure supplies to Richmond but not to Vicksburg. Otherwise one mirrored the other. Grant could not take Vicksburg by water only, Richmond by land only; he must combine the operations in both places.

Considering Pemberton on defense, we think first of Bazaine at Metz, MacMahon at Sedan, and Osman Pasha at Pleven.[21] They made the same mistake, Pemberton first. He had been so long so hard fortifying a place that he refused to leave. By remaining, he committed his fatal blunder. Wise general and farsighted strategist Joe Johnston said, advised, pleaded not to get encircled and cut off from the nation's resources. Keep in touch, stay in the grand design, be part of the armies in the field, he said. Pemberton disagreed. He would take advantage of Vicksburg's strengths. He thought he could defend those walls and hold the city with brilliance. Bravely he fought, tenaciously he held out, sadly he shared the fate of all who—like Bazaine, MacMahon, and the Osman Pasha—favor fortification and would save the country with fortresses. They must do what he had to do: surrender.

The Americans fought their civil war long ago and far from here. Yet we care what happened along the Mississippi then. The consequences resonate; they affect modern times; we feel them. We have seen combined operations work there, success when the army's troops and the navy's ships cooperated. In the history of modern war, as my readers will agree, perhaps no other episode provides so many and such good examples. One by one we have set them forth, then pointed out their lessons. Let us look back now in a brief summing up.

Union commanders acknowledged soon the value of army/navy cooperation: as early as the fighting at Forts Henry and Donelson. Boats accepted artillery of huge caliber for those times, and maneuvered it with ease hitherto unknown. Boats took what would make horses and wheels stumble and struggle and spin and sink and mire down. Boats moved the loads easy and fast. Waterborne artillery in action changed position as readily as field artillery never could. Waterborne under fire moved out of range as simply as it adjusted distance and chose targets when firing: not men laboring to drag guns like millstones over terrain like washboard, but a few turns with steam of screw or paddlewheel: simple, deft, quick. The army thus used the navy first and foremost as artillery support. In addition at the war's start the commanders valued steamboats as transports. They brought supplies and munitions up where needed, and here and there, moved troops as well.

Yet until the capture of New Madrid and Island Number 10, the navy remained secondary, the aide. By then, however, as gunboat construction got better, so gunboat support had become essential, indispensable, to army operations. At the Island, commanders learned the capability of the gunboat. Outfitted right, it could run in relative safety the strongest of forts. Better manufacture and safe mobility boosted "fighter-boat" stock on the commanders' exchange, and with it the commander's calls for gunboat cooperation in land operations. Therefore so early as the first strike at Vicksburg, Sherman relied readily, securely, and freely on the navy as transportation and a combat arm.

No overland mode could put more than the pace and the power of the snail against the boats' speed in conveying over untold miles the infantry's every man jack, the artillery and its equipment down to the final fuse and last bolt, and the cavalry and its every single horse. Boats moved these forces with ease and in comfort unknown on land. They arrived, not enervated, not tired, but battle-ready and near the objective. Could there be another advantage? Yes! At Vicksburg, not only the troops from Cairo safe and fresh after 450 miles, but with them their ammunition and provisions, all in never a whisper of danger.

Such feats could inspire awe that would excite overestimates of what the navy could do—of course. Sure enough, the navy delivered and the overestimates followed. Grant himself made the mistake, what with his trials and tribulations overland. He had set out, but Confederate cavalry—alert, busy, nimble, enterprising—soon cut his rail lines, seized or destroyed his stores, and defeated or pinned down his detachments. Nor could he operate in secret on terrain home to those horsemen, let alone succeed against their excellence in small-scale, intensive warfare. So he began to regard the congenial decks of Mississippi riverboats as havens of safety and succor. He would trust them to move the troops henceforth. Soon he relied on them. In the end he fell in love with them, an affair that blinded him. Thus—to stress the point, I repeat—the army overestimated the navy, depended too much on it in amphibious operations, and paid the price.

The Mississippi valley taught the army the navy's role in land warfare; what I have just said about Grant shows it. The navy played its role in that theater, after the army learned how much the navy could help, and how to use the help.

In every battle from Fort Henry to Grand Gulf the Union navy showed the Confederate what mobility meant. The Union also proved superior in size and build of guns. Boats supported the corps and regiments. Yet unless combined with guns and men ashore—unless behind field artillery and troops attacking on foot—waterborne artillery and naval thrusts would fail against an enemy on land. Army and navy learned their lessons in a harsh laboratory, that grim classroom of the valley, where they must exert themselves and suffer pain to pass the course. Men sweltered in Mississippi heat. Mosquitoes harassed them to rage. Swamps and lowlands bred fever. Work produced nought. Sickness flourished. Battles gained nothing. Failure for so long, bloody failure! At last, under the worst conditions and after supreme and exhausting efforts, troops landed on the left bank, back where they belonged, ashore. Only then, lessons learned, did Union forces get what they came for. They took the high ground and captured Vicksburg. And, secure there, they continued to succeed with a vengeance. The army, tried and tested for months, had learned the lessons of combined operations. The navy could help—it would be indispensable to campaigns and prove itself the staunch ally in battle—but campaigns must be conducted and battles fought by the army and on land: by it and no other service, there and nowhere else.

We close by restating the point. Let it be thus emphasized what the struggle for the Mississippi teaches by superb example. Strength and effectiveness follow interservice cooperation. Success and victory come with coordination, interaction, collaboration, and teamwork between army and navy in
COMBINED OPERATIONS.

# Epilogue

## The Lesson of the War

*Let us learn from the American war. It teaches the value of rectitude.*

### What Morality Meant to the Armed Forces—A Look at European Conditions—Examples.

No war in modern times has been fought from start to finish with the scales so tipped. Compare the armies in approximate numbers from 1861 through 1864. In 1861 the North had twice the troops. Lincoln called up 70,000, Davis 32,000. The ratio holds for 1862, except with Jackson in the Shenandoah. He must prevail against triple his numbers. In 1863 his became his country's odds. Consider the year's chief clash, the three days at Chancellorsville. Official reports give Hooker 159,000. I know by personal experience the Confederate numbers, being there watching from the first shot to the last: scarcely more than 50,000. In 1864, as I have suggested in the overview, Lee in Virginia with 81,000 had to struggle against 294,000. And while the North not only added men but also kept replacing losses with fresh men, Lee's band dwindled and shrank, shrank and dwindled, reinforced not at all. The venerable commander must contend with four times his number and even more later. Yet had Sherman not marched through Georgia, Lee so succeeded that the outcome of the war would have remained in doubt. Lee's South won nearly every battle in four years of war.

The numbers speak louder than words; they demand that the military analyst explain, especially establish the causes. How could armies of such different sizes be nearly equal in strength? Too often the causes have been distorted in what has been written; the authors have treated the numbers wrong. They have disrespected them and oversimplified the problem. They have cast the formula (a + b = c, or 2 + 2 = 4), and stopped when nonformulaic advantages and disadvantages ought to be factored in.

Seeking the causes, we discount heritage. Englishmen transplanted in North America founded and fathered the nation. They count on both sides. Again, in the final year, the sides enjoyed equal leadership and enforced equal discipline. In other words, in the year of the most disparate numbers, both sides had good commanders, and neither let discipline lapse. Neutral observers agree on discipline in the Northern then: excellent. On the other side one must notice that the Southerner grew up outdoors. Knowing the horse and the gun before becoming a soldier, he proved better prepared. The Northerner offset that advantage with technology: superior artillery, excellent small arms, and the navy that cut off essentials to the South as well as penetrated via rivers and streams to the Confederacy's heart. The South's mastery of basic soldiery could not diminish those strengths.

Then how can I claim Southern preeminence? Remember, I refer chiefly to the Army of Northern Virginia, the one that fought to the end and that I saw with my own eyes. And when I say preeminent, I mean superior in rectitude. Search and cogitate, you shall not find it otherwise nor discover anything else that so stiffened moral and ethical fiber man for man until the army could defy the enemy's advantages. Fewer but nobler nearly wiped out more but baser men.

The moral factor, the one to affirm in every generation, must be stressed now. Today's world ignores the essential of conscience and reveres egotism and wants to scorn and ridicule to death the desire for upright conduct. Materialism threatens the army itself, easy prey these days. Therefore the army must follow the principle that when asked for the supreme sacrifice, the soldier without religion being unfit, the officer without ideals being unworthy, they cannot measure up.

Even some military literature exudes today's miasma. Count among its faults the phrasemongery that ran riot a while back. If publications have had a wholesome effect of late, it has been by systematic examination of military topics. Scientific treatment has driven the phrasemongery out, swept aside bias-ridden practices, cast off conventions, and thrown light into one dark corner after another. But the abstractions of science can go too far. When we say warfare, we should mean art, not science. True, in our latest campaigns we advanced and won by complicated moves planned step by step. Astounding success tempted many of us to go for science. The impartial observer, however, saw the god of war overlooking battles, lifting his finger, warning: morality means more than calculation, rectitude more than reason. Let us learn from the American war. It teaches the value of rectitude.

I, a young officer who kept a diary, recorded what impressed me first and most in the Confederate forces. Entries speak often of the admirable attitude and inner harmony of the troops. These things caught my attention, struck me, because they contradicted the newspapers back home. What I saw belied the correspondents' sad reports. The longer with that army the more I learned to think and feel and—on days when cares weighed heavily—to consider myself at one with those men, thus higher and higher my regard for them. Lee ennobled whatever he looked at. To be in his presence compensated for hardship. A conversation with him, there amid civil war, lifted one above workaday concerns, even above worries about a continent in crisis.

Let me exhibit his order to the army in 1863. Civil war had hit the lush, prosperous Shenandoah valley with a vengeance. The army on the second grand expedition, marching north, must look at the havoc from morning until night. Hatred in the ranks ripened like corn, row upon row, a bumper crop. The army prepared to vent it and avenge the damage by wreaking their own on the other side of the Potomac. Lee's directive, General Order Number 73, June 1863, will show what I mean by his high-mindedness and Confederate rectitude.[1]

Headquarters Army of Northern Virginia,
Chambersburg, Pa., June 27.

The commanding general has observed with marked satisfaction the conduct of the troops on the march, and confidently anticipates results commensurate with the high spirit they have manifested. No troops could have displayed greater fortitude or performed better the arduous marches of the past ten days. Their conduct in other respects has, with few exceptions, been in keeping with their character as soldiers, and entitles them to approbation and praise.

There have, however, been instances of forgetfulness on the part of some that they have [an obligation] in keeping the yet unsullied reputation of the army, and that the duties exacted of us by civilization and Christianity are not less obligatory in the country of the enemy than in our own. The Commanding General considers that no greater disgrace could befall the army and through it our whole people, than the perpetration of the barbarous outrages upon the innocent and defenseless, and the wanton destruction of private property, that have marked the course of the enemy in our own country. Such proceedings not only disgrace the perpetrators and all connected with them, but are subversive of the discipline and efficiency of the army, and destructive of the ends of our present movement. It must be remembered that we make war only upon armed men, and that we cannot take vengeance for the wrongs our people have suffered without lowering ourselves in the eyes of all whose abhorrence has been excited by the atrocities of the enemy, and offending against Him to whom vengeance belongeth, and without whose favor and support our efforts must all prove in vain.

The Commanding General therefore earnestly exhorts the troops to abstain with most scrupulous care from unnecessary or wanton injury to private property; and he enjoins upon all officers to arrest and bring to summary punishment all who shall in any way offend against the orders on this subject.

R. E. Lee,
General.

I say here what I in accord with other foreign observers made public about that high-minded directive. As Lee admonished, so the troops behaved: with honor. True, one instance does not verify an army's morals. But such self-control must signify virtue. A second instance corroborates it. The army, without beer or wine as a substitute, gave up whiskey of its own free will. One unit's choice became the rule army-wide. Generals took energetic measures to enforce it.

Let us look closer at the troops. North versus South, which fought better and why? The South's fought like a people's army for the existence of the nation, for independence. I dare say they fought to preserve hearth and home. Indeed, they sensed what the history of revolutions shows, and they divined what the outcome of this war emblazoned: the defeated minority suffers. Furthermore the Southern soldier grew up not only in the free outdoors but also in a disciplined household, in a family with rectitude. Upbringing made him physically, mentally, and morally fitter than Yankees reared in cities and shaped by nerve-jangling hubbub. Manpower in two theaters illustrates what I mean. In the West the armies of North and South, men of similar childhood, differed in quality but little. Not so in the East. Southerners proved better that Northerners there: the sons of Virginia, Georgia, and the Carolinas over the offspring of New England. Finally, religion. (Scoff, you of the younger generation! Scoff, you who will! I say in spite of you: *religion*!) The Southern soldier's religious character sprang from his ethical and moral depths and helped make him braver and more dutiful. The old truth proved true again: "Good Christian, good soldier!"

To explain. The man happiest with life's pleasure clings with greater love to life itself. Fear of punishment may move him to noble acts. Dedicated to higher things, however, the man realizes his best by self-discipline and self-control in behalf of higher things. Him we find among the war's top commanders. Lee, Jackson, and Stuart put their Christianity first. The Southern officer, a leader like Cromwell and Gustavus Adolphus, served as priestly example to his troops. I do not mean an example by outward signs and hypocritical words. This officer sets the example of authenticity by severity in personal conduct and the abiding faith in Providence. How the men therefore trusted, indeed how they loved Lee and Jackson! With that love and an affectionate trust, how the men looked up to them as models without equal! To understand, let alone appreciate my last two sentences, you would have to live with that army as I lived with it.

Discipline in Lee's brave band accordingly never lapsed. With courage untainted by fear, they had repelled every assault to the last man. Driven at last out of fortifications around Richmond, victims of hunger, reduced to a mere eight thousand, surrounded in the wilds by a force ten times their size, the veterans rallied around the general as around a Higher Presence. They revered him. Finally, squeezed by the enemy's ring upon tightening ring, they must give up, and their old man must give them up with a heavy heart. How touching their surrender, how noble their end! To ease the difficult hour, they crowded around to touch, to caress him. The magnificence of surrender equaled the grandeur of battle. The war had ended, an army would end, but the love of men for officer prevailed. Even the enemy felt the sublime in the moment. A band of heroes surrendered; the enemy accepted with respect, with awe.

An officer may be ever so capable and energetic, but if he hold advancement as the object of his competence and effort, he shall prove unfit for it. Southern officers did not practice the so-called careerism. The office dedicated his every power to his country's freedom. This supreme, impersonal exertion smothered personal, petty ambition. The Cause came before the self. Indeed the Southerner, soldier as well as officer, championed the Cause. Neither knew envy nor felt ill to the other. Selflessness ruled. This sharing of one purpose and partaking in common in a noble endeavor produced common trust; amid ever-present danger it became respect; and trust and respect must lead to harmony among wills. Hence a congenial, unrestrained obedience to those high-minded, proven officers; nothing less than a vying by the men to do their duty; and among officers an absolute, unquestioning dedication in the face of peril to the Cause.

Serving in Lee's headquarters, I often saw Lee and Jackson communicate. In communiqués and letters and with news and advice they showed how Southern officers cheered and inspired one another. From those messages selflessness and cooperation shine with that army's heartening and productive spirit. The absence of careerism imparted charm as well as excellence to their commands. Do not miss the contrast to the selfish striving and the jealousy that fragmented the Northern, especially at first. Grant and Sherman trusted one another and shared each other's confidence, true. But their army did not win battles until they awakened the same in their men, ennobled them to sacrifice, requested the ultimate sacrifice, and saw it freely given. If martial impulse inspired victory, high-minded sacrifice won the war.

An incident from Longstreet's headquarters will illustrate Confederate selflessness. The commissary's clerk saw to the provisioning because it had overwhelmed the commissary, a major. The major did the clerk's paperwork in turn. One day the major went to the general's tent and confessed. "I'm incompetent. If the troops must eat, relieve me. Put the clerk in my place." The general did so. By choice the major reverted to private. The clerk's work had become routine with the major, over the years. The ex-major did not think it beneath him to continue it in the lowest rank. Need I say that this diminution of self brought forth categorical and boundless respect throughout his command?

One step often separates the noble from the base. When plunder and booty as always corrupt an army, the army finds the step easy, down from heroism to murder and arson. The army's morals never degenerate more than when the troops plunder by sanction. The army loses its self-respect, the depravity produces weakness, and the army destroys itself. The history of war asserts thousands of examples. Let me but allude to the end of Tilly after the rape of Magdeburg, and the collapse of Napoleon's army after the burning of Moscow.[2]

In the American war in the West, neither side shrank from brutality; both fought like savages. Morgan's raiders equaled the Union's hordes in wreaking havoc. In the East, Lee's Army of Northern

Virginia remained virtuous, lapsing but once. (I learned under Lee to expect discipline as a fact.) The lapse occurred only in a detachment. One of Early's, thrusting north of Washington, saw how the last Union campaign left the lovely, charming Shenandoah raped and desolate by burning and pillage. Enraged, Early's men wanted revenge and scorned restraint. Advancing through Union territory, they took revenge with fury beyond excuse. They fought bravely against all odds afterward, but nemesis soon visited. Sherman attacked and wiped them out.[3]

Of course we hear of the enemy as evil, monstrous. Too readily we believe the exaggeration. I do not want thus to degrade the Union army that opposed us in the East. But after poring over the many studies of this complicated war, and considering what the likes of Butler, Hunter, Milroy, Blenker, and Sheridan did, I must call their behavior what every other unbiased author has called it: wrong. Sherman's march through Georgia lost its luster when I read D. Coyningham, a Union captain.[4] He says, "War is very pleasant when attended by little fighting, and good living at the expense of the enemy.

"To draw a line between stealing and taking or appropriating everything for the subsistence of an army would puzzle the nicest casuist. Such little freaks as taking the last chicken, the last pound of meal, the last bit of bacon, and the only remaining scraggy cow, from a poor woman and her flock of children, black or white not considered, came under the order of legitimate business. Even crockery, bed-covering, or cloths, were fair spoils. As for plate, or jewelry, or watches, these were things rebels had no use for. They might possibly convert them into gold, and thus enrich the Confederate treasury.

"Men with pockets plethoric with silver and gold coin; soldiers sinking under the weight of plate and fine bedding materials; lean mules and horses, with the richest trappings of Brussels carpets, and hangings of fine chenille; negro wenches, particularly good-looking ones, decked in satin and silks, and sporting diamond ornaments; officers with sparkling rings, that would set Tiffany in raptures—gave color to the stories of hanging up or fleshing an 'old cuss,' to make him shell out.

"A planter's house was overrun in a jiffy; boxes, drawers, and escritoirs were ransacked with a laudable zeal, and emptied of their contents. If the spoils were ample, the depredators were satisfied, and went off in peace; if not, everything was torn and destroyed, and most likely the owner was tickled with sharp bayonets into a confession where he had his treasures hid. If he escaped, and was hiding in a thicket, this was *prima facie* evidence that he was a skulking rebel; and most likely some ruffian, in his zeal to get rid of such vipers, gave him a dose of lead, which cured him of his Secesh tendencies."

Coyningham continues in that vein—too much plunder and savagery to repeat here. I must say however that other authors related the same details—Major Nichols, for example, in the history of that great march.[5] Yet they provoked no outcry in the North, no desire for disarmament. People accepted them as nothing out of the ordinary, merely the obvious and self-evident facts of war.

Compare the introduction to the report of the committee on the conduct of the war. A pathetic statement, almost ludicrous, after what I have just quoted. The legislators say, for instance, with a hint of self-righteousness: "The members of the committee have continued the deliberations until the rebellion has been overthrown, the so-called confederate government been made a thing of the past, and the chief of that treasonable organization is a proclaimed felon in the hands of our authorities. And soon the military and naval forces, whose deeds have been the subjects of our inquiry, will return to the ways of peace and the pursuits of civil life, from which they have been called for a time by the danger which threatened their country. Yet while we welcome those brave veterans on their return from fields made historical by their gallant achievements, our joy is saddened as we view their thinned ranks and reflect that tens of thousands, as brave as they, have fallen victims to that savage and infernal spirit which actuated those who spared not the prisoners at their mercy, who sought by midnight arson to destroy hundreds of defenseless women and children, and who hesitated not to resort to means and to commit acts so horrible that the nations of the earth stand aghast as they are told what has been done. It is a matter for congratulation that, notwithstanding the greatest provocations to pursue a different course, our authorities have ever treated their prisoners humanely and generously, and have in all respects conducted this contest according to the rules of the most civilized warfare."[6]

True, the war would end when the Army of Northern Virginia surrendered—not before. The question therefore becomes much more important. Did Sherman have to destroy so much? His campaign ended the war. His barbaric holocaust did not shorten it by one day.

Lieutenant Colonel Henry Charles Fletcher, otherwise of the Scots Fusilier Guards, served on McClellan's staff in the Peninsula campaign. Fletcher did his best to be objective. Writing of the campaign, describing an army at plunder, he clinches my point.[7] Such an army cannot be restrained. Furthermore, its ability to fight suffers: so the record proves. The policies of Union generals McClellan and Meade illuminate the matter. In a well-run army they enforced discipline; their forces bear not the onus of guilt laid upon others.

Has it seemed bizarre that after our recent campaigns over here, I stress morality about a war over there and years back? I exaggerate? Army morals should be of indifferent concern? Look, then, if you will, at the life of General von Brandt.[8] Its lapidary prose drives the facts home. Why the ghastly French defeat in 1812? Neither the cold nor other factors defeated the French. They lost because of French selfishness and French greed for plunder. My remarks warrant an exclamation point, you say? Some of you, my fellow officers, will accept the reason, you who understand more of the army and better appreciate the times.

One more word on morality in the American war and how it applies to us. Do not take my remarks as the critique of a nation. The North not only cared like an angel of mercy for the sick: it also proved its virtue over and over. I limit my remarks to the army alone. I want to show that the absence of morality weakens an army—I intend nothing else. Of course the army's immorality and amorality must be ended where the army's ethics originate: in the nation and the educational system. The army can do little about morality, having to cope with a nation's folkways. Still the army must do all it can on its own and with its own. Let the army act thus and so. Let there be, not mere homily but a good example by the officer corps. Let officers in peacetime be upright and stress ethics, in wartime enforce the strictest discipline and inflict for the slightest infraction the severest punishment. I am not mouthing phrases, though we may see them often in books and pamphlets. I am stating principles. A moral code of rectitude, together with a vital Christianity, made the Confederate army a match for the enemy three times its size. The Confederacy's sources of strength—principles, rectitude, Christianity—have been and will be ours. They have brought honor to our Prussian name, and victory to our Prussian flag.

# Appendix
## The War in the West

*Thus ended naval maneuvers by Forrest's famous cavalry.*

### A Cavalry Escapade on the Water: An Interesting Moment in Combined Operations—Facts and Figures on the Armies and Navies.

We cannot call it quits yet; we have to mention an interesting moment in combined operations, the time the cavalry rode the waves. In this caper, horsemen (brave riders) seized and manned enemy ships, took them out to fight, and sailed to defeat.

In October 1864, Forrest got the order to strike again in Tennessee and block communications on the Tennessee River. Therefore General Abraham Buford's division of Forrest's cavalry went down the Big Sandy to its mouth on the Tennessee, south of Fort Henry a few miles. Fort Heiman and Paris Landing, old Confederate works, commanded the river there. Those works had been so well sited that they commanded it for two and a half miles. The forces disposed themselves: General Tyree H. Bell's brigade and part of a battery at Paris Landing, and General Hylan B. Lyon's at Fort Heiman. Lyon brought twenty-four–pound Parrotts and set them up. Downriver he put a few lighter pieces. Artillery had been ordered to hide and stay hidden from everything on the river until it approached within range, certain range. Staff officers with detachments scouted above and below, to observe enemy movement. The rest of the cavalry and Forrest himself arrived at Paris Landing on the Big Sandy October 29. The Union steam transport *Mezeppa*, full of matériel, and with her a barge, came within range the same morning, certain range. The artillery manifested itself, battered them, drove them to the shore opposite, and sent the crews scurrying inland. A captain of horse left Forrest's troopers, swam the Tennessee to the grounded craft, and brought back a small boat. With it the cavalry took possession of the *Mezeppa*, then, aided by an expert ferryman, towed her to their shore, and landed the valuable cargo, to their fortune.

Three Union gunboats appeared and started shooting. Forrest's artillery could not reach them. The boats hammered the Confederate position, forcing Buford to burn the *Mezeppa* and try to move the salvage inland. In the evening the gunboats withdrew. In the morning, October 30, the transport *Anna* appeared. The brave cavalry's artillery pounded her. Buford, wanting her undamaged, rode to the water's edge and shouted to the captain: Surrender. The steersman, as if to describe a semicircle to shore, kept turning and described a circle: back downriver. The artillery fired as much and as fast as it could, to no avail, as the target raced out of the ambush. In a few hours the gunboat *Undine,* escorting the transport *Venus* pulling barges, came like their predecessors downstream into sight. The Confederates waited until this prey not only appeared in the sights but approached nearly point-blank, certain range. The upper battery opened on the *Undine.* She fired back. They fought for an hour. Sharpshooters ashore, and the three-inch rifled cannon, forced her off and farther downstream and into Fort Heiman's range. She and those with her dared not cross that killing water. They turned back and into a bend of the river and its shelter upstream. The *Undine* kept shooting at the fort while repairing what the Union artillery had done to her. The steamer *J. W. Chaseman* took the turn now, and a pounding. Damage forced the *Chaseman* to shore at Paris Landing and into Confederate hands.

Soon after having learned of the capture and destruction of the *Mezeppa,* Forrest himself joined the action. General James R. Chalmers's division arrived the same afternoon. Next morning early, as ordered, Chalmers took post at the landing with a brigade and four guns. Buford meanwhile directed the battery there to close in on the *Undine* and the *Venus* and renew the attack. Colonel Becker reconnoitered, then followed with two regiments and two ten-inch guns. He ordered the cavalry off their horses. They crept through the underbrush to shore and opened rapid fire on the *Undine* and the *Venus*. The troopers and their small arms, shooting at the *Undine,* aimed at the gunports. The ten-inchers meanwhile took a position favorable to *their* fire. The small arms forced the ports shut. The ten-inchers scored so much that she had to grope to shore blind and crippled. Those of the crew not hurt abandoned ship for solid ground to save their skins. The *Venus,* steered to shore by now, had been secured. Two squadrons of cavalry-without-horse brought her and the barges back to Paris Landing.

The roar of artillery had attracted another gunboat. This one, taking up the fight, anchored about a mile upstream and started shelling the artillery ashore. The cavalry's arms could reach nowhere near the boat, of course. Hence no answer—yet. A detachment sneaked up and, sited afresh, poured fire on the new opponent. The newcomer had to get up steam, weigh anchor, and retreat upriver.

The captors destroyed the empty barges. Unable to refloat the *Chaseman,* they unburdened her of edibles—how they prized those delicacies!—stripped her of everything they could move, equipment and fittings too, distributed the haul among locals, and burned the rest. The *Venus,* one of the biggest gunboats, and the *Undine,* the lightly armored sister, needed no less than work by the corps' engineers, the machinery had been so badly damaged.[1] The engineers, detached for the purpose, did not finish until the evening of October 31. Work continued, arming the boats with eight twenty-four-pounders each, and otherwise readying them for combat. Forrest ordered the pair to fly the Confederate flag. He also assigned permanent crews, with cavalry captain Gracy to command the *Undine,* Lieutenant Colonel Dawson the *Venus,* and Dawson the head of the fleet. Forrest came aboard this new navy of his and ordered the boats to Fort Heiman on a shakedown cruise. They rounded the bend, nearly there. The units on land had lined up along the bank: every last trooper. From them rose cheer upon cheer and hats into the air in the ovation for the "cavalry of the waves."

The joy unfortunately died soon. Forrest ordered the boats downstream and, ever-cautious, sent artillery besides, overland. Bad roads and rain slowed this escort, nearly stalled it. The boats went ahead then, dangerously too far, rushing to their fate. Suddenly they faced, they took on, but they could not match three gunboats. The Federals' barrage kept the *Undine* off while the Confederates abandoned the *Venus* to her original owners. The *Undine* stayed Confederate, taking cover under the guns of her escort ashore. The Union's gunboats dared not follow. Other Confederate batteries, brought in to redress the balance against the gunboats, bombarded them from upstream and down. Can we know what the outcome would have been had the affair ended then and there? It ended there but not yet. More enemy appeared; three became eight. Onto the cavalry and artillery their big guns poured such fire that horses bolted from blast after blast and explosion upon explosion, and in panic galloped like crazy and carried their riders through underbrush toward safety inland. The *Undine* proved as incapable as her shoreside allies. Her impromptu sailors, used to the high horse and wide-open spaces, felt more than ill-at-ease in her close quarters, worse than discomfitted in those constricted casemates. Three hits soured them on seafaring one-two-three forever. They "let 'er rip," rammed the shore at full throttle, set fire to the decks, took horse at breakneck speed, and tore out of there, swearing never, but never, to play sailor again.

Thus ended naval maneuvers by Forrest's famous cavalry.

## 2

I. Orders of Battle: Army
  A. Union, General U. S. Grant, total about 70,300 at Vicksburg.
    1. Thirteenth Corps, General John A. McClernand
        a. 9th division,     Osterhaus,     6,200 men
        b. 10th                Smith            6,500
        c. 12th                Howey           7,000
        d. 13th                Ross             4,700
        e. 14th                Carr             4,700
        f. 2nd cavalry     Bussey          4,700
                                                  33,800
    Fifteenth Corps, General William T. Sherman
        a. 11th Division,   Steele          8,300
        b. 5th                 Belair           6,900
        c. 8th                 Tuttle           4,000
                                                  19,200
    Seventeenth Corps, General James B. McPherson
        a. 3rd Division,    Logan           7,000
        b. 6th                 McArthur      4,600
        c. 7th                 Quimby        5,800
                                                  17,400
    Sixteenth Corps, General Stephen A. Hurlbut, divided between Tennessee and Kentucky, thus partial participation, 62,000 men.
  Confederacy, General John C. Pemberton, Spring 1863
    1. Stevenson's Division,     33,000 men[2]
    2. Smith's                        7,100

          3. Forney's                4,200
          4. Boben's                 5,500
          5. Loring's                6,200
                                    39,300

   *Note.*—Pemberton could also draw on the Third District, a part
   of his jurisdiction and under General Franklin Gardner at Port
   Hudson; 20,000 men.
II. The Fleet on Western Waters, 1862: Union
    A. Commandant and staff. Command boat *Benton*
       1. Flag Officer, Admiral Andrew H. Foote, commanding
       2. Commander Pennock, commandant of Cairo
       3. Lieutenant Sandford, artillery
       4. Lieutenant Byron Wilson, artillery
       5. Flag Lieutenant Pritchet
    B. Boats and commanders
       1. *Benton.* Lieutenant Phelps
       2. *Carondelet.* Commander Walke
       3. *Essex.* Commander Porter
       4. *Louisville.* Commander Dove
       5. *Mound City.* Commander Kitty
       6. *Cincinnati.* Commander Stembel
       7. *Cairo.* Lieutenant Bryant
       8. *Pittsburg.* Lieutenant Thompson
       9. *St. Louis.* Lieutenant Paulding
      10. *Conestoga.* Lieutenant Phelps
      11. *Lexington.* Lieutenant Shirk
      12. *Tyler.* Lieutenant Givin

   *Note.*—The ordinary steam transports *Conestoga, Lexington,* and
   *Tyler* differed from the rest as warships. The three had gunports
   and perpendicular armored hulls, part of their refitting for war.[3]
   Worthless on the Mississippi, they excelled on tributaries with
   shallows posing risk to bigger boats of deeper draft. Except
   perhaps the *Benton,* bigger boats dare the Tennessee and the
   Cumberland only at high water.
    C. Construction of the Riverboat Fleet
       1. Merchant vessels converted to gunboats. (Most seagoing
          ironclads followed the lines of the *Monitor,* the famous
          example of the gunboat with the revolving turret. River-
          boats tended, on the other hand, to follow the Confederate
          entry in the notorious duel that struck terror into the

Union fleet. Why imitate the *Merrimac*, the boat of the enemy? She had been civilian, a merchant vessel; and for inland service the Union converted merchant vessels—passenger boats and freighters, civilians—already on the river.[4] Slanted casemates went over the decks and took sheets of iron as armor, producing something like a turtle's back sticking out of the water. Americans called them turtles. By and large they proved themselves in the Mississippi service; they could fight. This class included the *Cincinnati, St. Louis, Louisville, Cairo, Carondelet, Mound City,* and *Pittsburg.*)

Approximate specifications[5]:
  a. Length 55 meters, beam 24, draft 2, capacity about 500 tons
  b. Hull, vertical, under flanks rising 30 centimeters out of the water and sloping 45 degrees to the deck
  c. On deck a casemate housing machinery and cannon; 47 meters long, 24 wide, and 2.5 high on the inside; slanted as indicated above; of stout wood and armored with 6-centimeter iron plates
  d. Pilothouse, armored
  e. Power: usually sternwheel-driven at speeds to nine knots
  f. Armament: typically thirteen cannon, usually rifled. Example: *Carondelet,* three gunports at bow, two at stern, and four to either side; i.e., two rifled 42-pounders and one smoothbored 64 at bow; two rifled 32s at stern; one smoothbored 64, one rifled 42, and two rifled 32s at either side

2. Rams. Charles Ellet, designer; armed with but one howitzer; 200 to 400 tons; paddlewheelers[6]
3. Mortarboats. Simple barges converted; one 40-centimeter mortar per boat; towed, no propulsion
4. Wooden vessels.[7] Civilian riverboats bought for military use and converted; no armor; mostly 400 to 600 tons; about nine cannon; for example, *Tyler, Lexington, Conestoga,* and many others; ordinary steam freighters hastily converted to gunboats, provoking no imitators from then till now, therefore not worth description

III. The Fleet on Western Waters, Confederacy. The South built and improvised like the North. Their craft, however—some armored like the North's though with weaker plates, others with railroad rails—lacked basic structural strength against attack. Nor could the machinery be relied on, apt to fail at any moment. Repair the damaged? Build new? How hard in a South short of foundries and machine shops!

*Note.*—The Union navy too suffered inadequate vessels and defective equipment. We have it from Henry Walke, the *Carondelet*'s famous commander, that the construction of western waters' gunboats proved so defective as to pose danger after danger.[8] Random torpedoes could sink you, he said. One shot into the boiler or the powder magazine—neither of those vitals had enough armor—could blow you sky-high. Worse, the muzzles of guns being fixed too close to the decks, you could shoot at others and sink yourself. Crews below, stokers and engine men, thus found themselves often tending the fires of a boat on fire: fires all around. You had always to be on guard against fires, therefore, and carrying water to put them out, and at the same time ever-vigilant to the threat of being blown up.

# Notes

## Prologue: The Great Captains

1. Other Prussians, too, especially of the nobility, identified with the Southern aristocrats and their war to maintain a modern feudalism. Scheibert's friend Heros Von Borcke for example, the "romantic young giant" with the "Damascus sword" (Jay Luvaas, *The Military Legacy of the Civil War: The European Inheritance*, 56), fought for the South, excited Southerners as much as they excited him, and stayed in Southern memories. A century after the war, Southerners were still naming dogs and cats Heros Von Borcke. To have fought and bled for this glorious struggle, he said, gave him the greatest satisfaction of his life. Heros Von Borcke, "Memoirs of the Confederate War for Independence," *Blackwood's Edinburgh Magazine*, September 1865, 269.

2. Scheibert says *Observationscorps*. Observation, though not all the work at Harpers Ferry, had to be part of it: watching enemy movements and assessing civilian morale in the region. All in all, however, Jackson commanded the outpost, with orders to make soldiers of volunteers, construct defenses, and move the arms-fabrication machinery to a safer place, preferably Richmond. Frank E. Vandiver, *Mighty Stonewall*, 136.

3. Scheibert quotes the servant's words in English.

4. In point of fact, he seems to have said, "in delirium, 'order A. P. Hill to prepare for action!'" (Lenoir Chambers, *Stonewall Jackson*, 2:442).

5. August Graf Neithardt von Gneisenau (1760–1831), Prussia's strategist during the Napoleonic Wars, served as chief of staff to Gebhard Leberecht von Blücher (1742–1819), leader in Prussia's war of liberation against the French, 1813–1815.

6. Scheibert may be reporting what he heard from Lee's lips. See G. F. R. Henderson, *Stonewall Jackson and the American Civil War*, 692. Margaret Sanborn, *Robert E. Lee*, 2:114.

Scheibert has Lee's letter in full in German. Lee wrote (U.S. War Department, *The War of the Rebellion: A Compilation of the Official Records of the Union and Confederate Armies . . .*, ser. 1, vol. 25, pt. 2, p. 769):

> HEADQUARTERS,
> May 3, 1863.
> General THOMAS J. JACKSON,
> Commanding Corps:
>
> GENERAL: I have just received your note, informing me that you were wounded. I cannot express my regret at the occurrence. Could I have directed events, I should have chosen for the good of the country to be disabled in your stead.
> I congratulate you upon the victory, which is due to your skill and energy.
> Very respectfully, your obedient servant,
> R. E. LEE, General

7. F. von Meerheimb, *Shermans Feldzug in Georgien*, 11–12, 50–53. In the translation, two sentences of Scheibert's quotation have been deleted as contributing nothing to the sketch. Scheibert quotes as if presenting a complete, continuous passage, but the page numbers from Meerheimb show that it occurs in two places. In addition, Scheibert occasionally paraphrases rather than following Meerheimb faithfully.

8. Grant, drinking heavily, had neglected his duty and "resigned in 1854 to avoid court-martial" (Mark Mayo Boatner, *The Civil War Dictionary*, s.v. "Grant, Ulysses Simpson," 352). Indeed, he drank all his adult life. In 1854 "he did not leave the army because he was a drunk. He drank and left the army because he was profoundly depressed" (William S. McFeely, *Grant: A Biography*, 55). "At the beginning of the Civil War" Grant "would probably have been voted 'least likely to succeed'" (Boatner, *Civil War Dictionary*, 352).

9. That is, Grant took charge of the District of Ironton, Missouri, August 8, 1861, and Federal forces at Cape Girardeau, Missouri, in Jefferson County, locus of action in the West, September 1, 1861. E. B. Long and Barbara Long, *The Civil War Day by Day: An Almanac 1861–1865*, 106, 113. John S. Bowman, *The Civil War Day by Day*, 39. General John C. Frémont had commanded the U.S. Western Department since July, but Grant also owed his position to Congressman Elihu B. Washburn, who instigated Grant's promotion to brigadier general by President Lincoln, July 31, 1861. Boatner, *Civil War Dictionary*, s.v., "Frémont, John C.," "Grant, Ulysses Simpson," "Washburn, Elihu B."

Grant, without fighting, seized Paducah, Kentucky, his first victory, September 6, 1861, before General Leonidas Polk turned command in the West over to General Albert S. Johnston, September 15, 1861. Long, *Civil War Day by Day*, 115, 118.

10. Scheibert implies a President Grant as meritorious as the General Grant. Much of the politically incompetent Grant's ignorance as president had not appeared when Scheibert prepared this book.

11. That is, the English Civil War, 1642–1648. Scheibert implies that the Lees might have left because of its political troubles. Not so. Those Lees, with the energy and enterprise of every Lee generation, left for what immigration would gain them in the New World. See Paul C. Nagel, *The Lees of Virginia: Seven Generations of an American Family*, 9–13.

Robert E. Lee married the daughter of George Washington's adopted son and enjoyed Arlington, the part of the Washington estate she inherited. Lee "never owned a square foot of land" himself (Thomas L. Connelley, *The Marble Man: Robert E. Lee and His Image in American Society*, 7). Nor did he need to, being a military officer with government quarters, then head of Washington College, residence again provided. See Nagel, *Lees of Virginia*, 234, 235–99 (esp. 235–39, 286–89).

Scheibert says *politschen*. Lee's forbears, back to the first in America, had held public offices and been distinguished political leaders. Lee himself had been and would be a distinguished military leader only, until becoming a college president late in life.

12. That is, he could have led the armies of the Union but chose to serve Virginia and the Confederacy instead.

13. This paragraph may report what Freeman calls "a mistaken theory of the function of high command" (Douglas Southall Freeman, *R. E. Lee: A Biography*, 4:168; see also 2:347).

14. Supplies failed to arrive at Amelia Courthouse, forcing Lee to raid the countryside near the end of the Appomattox campaign.

15. That is, Washington and Lee University at Lexington, Virginia, then called Washington College. On Lee's discussion with Scheibert of Southern youth and their education, see Freeman, *Lee*, 2:538, 4:217.

16. "The first soldier of his country" suggests Frederick the Great as the object of this openly veiled encomium. "Respect for the living" and "loyalty as an officer" suggest Napoleon I—not to be mentioned by name by Scheibert, a loyal officer who fought and was wounded against Napoleon III in the Franco-Prussian War two or three years before writing these words. The reference remains unclear, the allusion unidentified. Still, the best guess may be Helmut von Moltke (1800–1891) field marshall and chief of the Prussian (and, later, German) general staff: Scheibert's superior at this time and for years before and after.

## Chapter 1. Strategy

1. "Plank roads" is in English in the original.

2. Scheibert implies that mobility has become sacred by 1874 in the German army, and everything impeding it heretical. Indeed, in 1870, Moltke's Prussians smashed Napoleon's French by mobility, maneuver, speed, surprise, and shock.

3. Johnston failed that day at Fair Oaks, nonetheless, in spite of his interior lines. Here and everywhere with Scheibert's numbers, pay attention to proportions or other points being made. His numbers of themselves, like most at that time, often approximate rather than tally with later calculations by historians and the government.

4. By "return south," Scheibert probably means that the Union leadership wanted to keep Jackson from reinforcing Lee around Richmond.

5. That is, Lee would now withdraw from the defense of Richmond against McClellan—without squaring off against McClellan again—to oppose Pope, and with Jackson and Longstreet, defeat Pope at Second Bull Run in late August 1862.

6. Scheibert refers to the light engagement at Malvern Hill and the skirmishes at White Oak Swamp and Thornburg or Massaponax Church.

7. In point of fact, significant action occurred in Kentucky, West Virginia, and Tennessee, not Missouri. Scheibert says nothing about the Battle of Antietam (Maryland). This would be the time and place in his narrative to acknowledge the battle "considered by some historians to be the turning point of the war" (Boatner, *Civil War Dictionary*, s.v., "Antietam [Sharpsburg] Campaign," 21).

8. Scheibert has "mud campaign" in English. Documents of the time often use another term, perhaps more usual: "mud march."

9. Scheibert refers to the Chancellorsville campaign, events leading up to the Battle of Chancellorsville, and General Pleasonton's testimony of March 7, 1864. "I went to General Slocum . . . and told him that the rebel army had not moved from Fredericksburg, and did not know of the crossing of [Hooker's] three corps at the Rapidan" (U.S. Congress, *Report of the Joint Committee on the Conduct of the War* 1:27). The phrase "big battle" is in English in the original. Regarding Hooker's crossing of the Rapidan, Scheibert in a note to the original at this point refers the reader to his *Seven Months in the Rebel States during the North American War, 1863*, on the Battle of Chancellorsville, 51–60. "Stuart was to observe the crossing of the enemy at Kelly's Ford" (52).

10. That is, laid in bravado but implemented with cowardice.

11. Abraham Lincoln, *Collected Works,* 6:249. Lincoln's language has been supplied, not a translation of Scheibert's translation. For "Hooker's plan" see editor Basler's note, pp. 249-50.

12. Scheibert, so describing the letter, says, in effect, that the cabinet's narrow-mindedness includes Lincoln himself. The reference to Austria may be to Austria's defeat in the Austro-Prussian War (1866) and to its decline thereafter, especially in relation to Prussia and Germany, Scheibert's countries.

13. U.S. War Department, *Official Records,* ser. 1, vol. 27, pt. 3, p. 70. In all references to Pleasonton in the rest of the paragraph (except the letter to Butterfield), Scheibert seems to be loosely paraphrasing Pleasonton's testimony of March 7, 1864, U.S. Congress, *Joint Committee Report,* 1:32-33.

14. For letter, see U.S. War Department, *Official Records,* ser. 1, vol. 27, pt. 3, pp. 70-71. In other words, Pleasonton reports success for himself and the Federals at Winchester and against the invasion of the North. In fact, Ewell and especially Rodes of Ewell's corps defeated the Federals there, and the Confederates began the successful crossing of the Potomac. See Long, *Civil War Day by Day,* 366, 367, 368, 369.

15. Scheibert refers to the June skirmishes and engagements at Middleburg (June 17), Aldie (June 18), and Upperville (June 21). (The reader should also bear in mind Scheibert's description and discussion of the Confederate army's move toward Gettysburg.) Von Borcke had taken a ball through the neck in the engagement at Aldie. See Scheibert, *Seven Months,* 104-8.

16. Scheibert refers to Missionary Ridge to mean the loss of Chattanooga, "a severe blow to the dying Confederate cause." Boatner, *Civil War Dictionary,* s.v., "Chattanooga Campaign," 147. Scheibert has the sequence of Chickamauga/Chattanooga wrong but makes the point nonetheless: troops from Virginia did reinforce Bragg at Chickamauga, and he won there.

17. Scheibert includes Petersburg in the study of tactical defense, rather than as an example of siege. He says in a note to the original here: "Lee did not hold a fortified site at Petersburg. Grant therefore did not besiege it. They fought a long, wearisome, tactical battle for it." By Scheibert's count, Grant had 194,700 men; Lee, 52,600.

The metaphor of chess, good enough in English, works better in German: *der Laufer* (bishop) being literally runner or sprinter; *der Springer* (knight), jumper or vaulter; and *der Turm* (rook), tower or castle.

18. This reference to Petersburg contradicts the one above—was Petersburg an example of tactical defense or of siege?—unless Scheibert restricts siege to operations against forts.

## Chapter 2. Tactics

Scheibert in this chapter limits his discussion to infantry tactics. Elsewhere in the book he touches upon the others.

1. "Let me mention a report out of that time, Colonel William Wilson's of Pensacola Bay, commander of Zouaves. It illustrates the tone of the many then" (Scheibert). Wilson's report has not been identified in the original. On later pages, Scheibert will speak of more European attention than he leads the reader to believe here. According to Scheibert, Wilson writes of five men and two women from Pensacola, who tell how—among the rest of the news—people

there speak of the problematic victory at Bull Run. Meanwhile, indeed, Wilson and his men have been going to bed armed: he has not slept much; he does not feel good; he has diarrhea; his men need eight hundred uniforms.

European interest, rather than decline, quickened and enlarged. More newspapers sent correspondents; armies, notably the British and the French, dispatched observers. If Prussian attention lapsed it must have been recovered, else why send Scheibert in 1863? European practice suggests that Europeans learned at once the American lessons of railroads, medical services, arms and equipment, and the telegraph and other communications. Prussia seemed to apply, at least did not contradict, them as early as the 1864 war with Denmark. But lessons of tactics, especially with respect to the new firepower, had not been learned as late as 1870. French and Prussians alike sent cavalry against infantry armed with breechloaders: bloodbaths; tight formations of infantry against rifled artillery: slaughter; and frontal attacks of infantry against breechloader-armed infantry, again, again, and again, yet again: carnage. Scheibert may have written his 1874 book against such ignorance of what ought to have been learned a decade before.

2. U.S. War Department, *Official Records*, ser. 1., vol. 2, pp. 321, 325. McDowell's language has been supplied, not a translation of Scheibert's translation.

3. Scheibert has *Kriegsartikel* (martial law) for what McClellan enforced. The term could mean, more or less literally translated, articles of war or even Scheibert's martial law. But neither term fits here. Hence I have used "regulations for the government of armies in the field" from U.S. War Department, *Official Records*, ser. 3, vol. 1, p. 395.

4. Helmut Karl Bernhard, Graf von Moltke (1800–1891), Prussian field marshal, molded the Prussian army and developed the tactics that won three wars in close succession, last being the Franco-Prussian (1870–1871), the one that created Scheibert's Germany of the time of this book (1874).

5. Stuart's cavalry had been sent around the Federal army on June 25, an action "over which controversies rage still." Rejoining Lee on July 2, the cavalry missed much of the Gettysburg operation and, as Scheibert says, could not be the eyes and ears to Lee's army. See Long, *Civil War Day by Day*, 371, 372, 377. The quoted phrase is on 371.

6. The usual figures put Grant's at about twice Lee's (not Scheibert's four times) or perhaps 110,000 to 60,000.

7. In the chapter on infantry, Scheibert speaks of the regiment as the tactical unit. He seems not to be contradicting himself, rather saying that in these special circumstances and perhaps in the regular army, the brigade could be regarded as the tactical unit. "Essentially, brigades did the fighting in the Civil War" (Richard J. Sommers in *Historical Times Illustrated Encyclopedia of the Civil War*, ed. Patricia Faust, s.v., "Tactics").

## Chapter 3. Infantry

1. Francis Parkman saw "the backwoods lawyer" as the officer in the army of 1846, "better fitted to conciliate the good will" than command the obedience of "many serving under him who both from character and education could have better held command than he" (*The Oregon Trail*, chap. 26).

2. U.S. Congress, *Joint Committee Report*, 1:444–46. Testimony by General John Gibbon, April 1, 1864. The pages contain the testimony quoted here by Scheibert

and Scheibert's paraphrase in the paragraph following the last quoted answer. I have given the language of the testimony, not my translation of Scheibert's translation. Scheibert gives "General Gibbon" and "General G." instead of the "Answer" of the transcript. I have italicized all three for easier reading. I have moved the quotation and the paraphrase to the place in the text from Scheibert's addendum at the end of his chapter.

3. Scheibert says below, correctly, that Union and Confederacy followed Casey's and Hardee's tactics, respectively, and that the two systems resembled one another more than they differed. One might read Casey as calling the brigade the unit, but the discussion that follows Casey's statement, seemingly Scheibert's source here, allows flexibility in names and numbers; and as Scheibert has, one can take either the regiment or the battalion, as well as the brigade, as the unit, depending on numbers and circumstances. Silas Casey, *Infantry Tactics for the Instruction, Exercise, and Manoeuvers of the Soldier, a Company, Line of Skirmishers, Battalion, Brigade, or Corps d'Armee*, 1:9–12. Throughout his three volumes, Casey refers more to brigades and battalions than to regiments. Scheibert here and in what follows in this chapter, despite his references to Casey and Hardee, may of course be speaking according to what he saw rather than what they say. As for resembling the French, Casey based his work on the "French *ordonnances* of 1831 and 1845." Casey, *Infantry Tactics*, 1:5. Jack Coggins, *Arms and Equipment of the Civil War*, 21–25, citing Arthur L. Wagner, *Organization and Tactics*, sounds more like Scheibert than Scheibert sounds like Casey or Hardee, suggesting that Scheibert followed other sources than them or wrote from observation as much as from documents.

4. Comprehensiveness, size, and date tell why they chose Casey's tactics: Casey, *Infantry Tactics*.

5. See William Joseph Hardee, *Rifle and Light Infantry Tactics for the Exercise and Maneuvers of Troops. . . .* Hardee, a Confederate officer, wrote for the U.S. Army before the war. As Scheibert says, Hardee and Casey duplicate one another in many ways. Hardee translated the *ordonnances* used by Casey in *Infantry Tactics*, 1:5, hence the French influence on both Casey and Hardee.

6. The formal term, *dismounted cavalry*, applied not only to Jackson's but also to cavalry of both sides throughout the war. They deserved the title, being more infantrymen who rode horses to the battle, and less (even seldom) troopers or dragoons who fought on horseback. See Grady McWhiney and Perry D. Jamieson, *Attack and Die: Civil War Tactics and the Southern Military Heritage*, 134–36.

## Chapter 4. Cavalry

1. The Crimean War (1853–1856) and the Austro-French-Sardinian War (1859). In "The Charge of the Light Brigade," Tennyson dramatized the futility of cavalry against entrenched firearms. At Balaclava, October 25, 1854, two-thirds of the brigade fell dead or wounded: "Theirs not to reason why / Theirs but to do or die."

2. Scheibert with this paragraph, as well as elsewhere, seems to qualify his assertion that Europeans lost interest after First Bull Run. Scheibert wrote to the editor of the *Southern Historical Society Papers* that Southern examples had been the pattern for Prussian and German, including "our cavalry tactics," which General Von Schmidt remodeled after asking Scheibert about Stuart's. *SHSP* 9 (1881): 571.

3. Charles Cornwallis Chesney (1826–1876) wrote *A Military View of Recent Campaigns in Virginia and Maryland.* An army officer, author of books, essays, and papers on military topics, and "the best military critic of his day," Chesney "subjected the American civil war to a close and searching military criticism" in that book "recognized as a valuable contribution to military history" (*Dictionary of National Biography*, s.v., "Chesney, Charles Cornwallis").

4. That is, the North, developing a different cavalry from the South's, fought with guns rather than shock and saber, and mounted as well as dismounted.

5. That is, in the interest of speed, they carried their provisions when speed did not matter. When it mattered, they for sustenance relied on units visited and must be served first to insure getting something and as quickly as possible.

6. Scheibert, using the word *scouts,* calls it untranslatable into German.

7. Scheibert has *hunter* and *trotter* in English in the original. Mosby's Partisan Rangers deviated from the pattern—see the paragraph above—they attacked on foot.

8. Scheibert neglects to recall that, as he noted in the chapter on tactics, the absence of cavalry hurt Lee at Gettysburg.

9. Such gestures by sharpshooters were used to make the enemy move, perhaps deploy, and thus reveal himself. Again (see note 7), Scheibert does not mention what he said of Mosby's Rangers. But rather than fight on foot, cavalry would run so fast as to abandon their own men.

10. Stuart's cavalry, though called a corps, remained a division in organization and size even after Chancellorsville. Reorganization after Gettysburg made it a true corps.

11. I quote not my translation of Scheibert's translation, but Von Borcke's English in the "Memoirs" (October 1865, 410–12, 414–15, 416–17). Scheibert's German occasionally deviates unimportantly from strict faith to the original. Scheibert explains by footnote the meaning of battle flag: "Each headquarters carries this marker in the field, a cloth about a meter square on a short staff, to identify the unit."

12. It began as the series in *Blackwood's,* later published as a book.

13. *Blackwood's,* September 1865, 281–82. What note 11 says about translation applies here.

14. Scheibert, an engineer, suggests in part here that cavalry might consider the utility of engineering and realize that engineering can be of value even to the extent that the cavalry needs engineers or troopers who can do engineers' work.

## Chapter 5. Artillery

1. True, both sides preferred the standard Napoleon, but it was smoothbored. See Boatner, *Civil War Dictionary,* s.v., "Napoleon Gun Howitzer"; Faust, *Encyclopedia,* s.v., "Napoleon." Scheibert in this sentence may have confused the smoothbored Napoleon with any of the several rifled twelve-pounders: the twelve-pound gun; or the three-inch ordnance gun, called a twelve-pounder though it fired a ten-pound projectile; or with the twelve-pound James, twelve-pound Blakely, or twelve-pound Whitworth, all rifled. See Faust, *Encyclopedia,* s.v., "Napoleon," 755, and s.v., "Three-inch Ordnance Gun"; Francis A. Lord, *Civil War Collector's Encyclopedia,* 24; Henry L. Scott, *Military Dictionary,* 66; Harold Leslie Peterson, *Notes on Ordnance of the Civil War,* tables on artillery, no

pagination. Scheibert's descriptions and specifications of artillery nearly always agree with Peterson's book and with Coggins, *Arms and Equipment*. Except this remark on the Napoleon, Scheibert's few and minor deviations deserve to be heard and could be correct. "Sources," Peterson says in *Notes on Ordnance*, "frequently disagree with each other because of variations in individual guns and in the degree of accuracy . . . of tests of velocity and range."

2. With *pace* I have retained Scheibert's use of *Schritt* (step, stride, pace) for ranges. It probably signifies something less than a yard and therefore cannot be translated for sure. *Pace* retains fidelity and authenticity: the Civil War artillerist in combat likely would have measured or estimated in the readiest way to determine ranges. Old-timers in field artillery still say *pace*.

3. Numbers for organization and accessories, like those for ordnance specifications, vary from author to author. Scheibert's term for the carriages, *Blocklaffeten*, translates literally "block carriages," the term originally for those "made from a solid piece of timber, hollowed out so as to receive the gun or howitzer" and used "in the English artillery," Louis de Tousard, *American Artillerist's Companion*, 2:453, also called the English block-trail system. William E. Birkheimer, *Historical Sketch of . . . the Artillery, United States Army*, 273-89. The American carriage, however, usually consisted of two wheels and the barrel and firing apparatus on the axle, held in place by cheeks on either side of the barrel. A stock or tongue extended to the rear, for drawing by horses and wheeling and turning by cannoneers. The machine could be called the cheek-and-block wheeled carriage. Scheibert may be referring to it rather than the English block carriage; for the cheeks' shape approximated rectangular blocks, suggesting objects referred to as blocks (*Laffeten*) in German. See Tousard, *Companion*, plate volume, plate 20. Scott, *Dictionary*, 150, s.v., "Carriages"; John Gibbon, *The Artillerist's Manual*, 180-81, 186-87; U.S. War Department Artillery Officers' Board, *Instructions for Field Artillery*, 128 and plate 8 following 139.

4. Scheibert does not identify the author.

5. Scheibert seems to have confused two writers: Prince Kraft Karl August Hohenlohe-Ingelfingen (1827-1892), author of many works, including some on artillery, and R. Roerdanz, *Das gezogene vierpfundige Feldgeschütz* (The Rifled Four-Pound Field Piece) (Berlin: Mittler, 1864), 55 pp., a lecture published as a pamphlet. His own writings on the topic include *Einfluss der neuesten Taktik und der gezogene Waffen auf den Festungskrieg* (The Latest Tactics, Rifled Ordnance, and Their Influence on Siege Warfare) (Berlin: Mittler, 1861).

Scheibert, assigned to observe the effects of artillery upon fortifications, would need the knowledge and skill of the engineer and artillerist.

6. Scheibert calls the place the Trafalgar Works.

7. Scheibert refers either to a battery under Cheves's command or to the artillery of Morris Island, Langdon Cheves, engineer-in-charge. Compare Long, *Civil War Day by Day*, 409, and U.S. War Department, *Official Records*, ser. 1, vol. 14, p. 958.

8. S. Jacobi, *Die gezogenen Geschütze der Amerikaner bei der Belagerung von Charleston* (American Rifled Artillery at the Siege of Charleston) (Berlin: Vossische Buchhandlung, 1866). For Charlestonians' experience of such artillery—the nature and strength of the defenses, those guns' misfires or blowing themselves up, the horror of attacking missiles' exploding in the streets or bursting on roofs, and the flying and falling debris and the fires: descriptions to send chills up and down the spine even now—see Scheibert, *Seven Months*, 132-41; Frank Vizetelly,

"Charleston under Fire," *Cornhill Magazine* 10 (July 1864): 90–110; and "The Siege of Charleston," *Harper's Weekly* (January 9, 1864): 17, and the sketch "by an English artist," of a bomb bursting in a street (28).

9. In the original, Scheibert placed a question mark at the end of this sentence, as if to doubt the wisdom of the process or the accuracy of the statement.

10. Colonel Alfred Rhett, First South Carolina Artillery, commanded at Sumter. See his reports of the bombardment. U.S. War Department, *Official Records*, ser. 1, vol. 28, pt. 1, pp. 608–22.

## Chapter 6. Engineering

1. Scheibert's "pedantry of systems" must refer to theories of fortification calling for shapes to protect flanks, usually a polygon exact in angles, each side having "due proportion according to rule." See Scott, *Dictionary*, s.v., "Fortification," 316–17.

2. The city, besieged by the British, French, Turks, and Sardinians in the Crimean War, held out for nearly a year, thanks to the ingenuity and strength of fortifications. A year under siege probably caused plenty of the nervous fatigue that Scheibert wants to reduce.

3. *Torpedo* in Civil War days meant what we call the mine: both land and marine. The marine might be set in motion to a target, rather than placed to let the target do the hitting, hence today's understanding of torpedo. Scheibert also mentions, below, the floating rocket, another predecessor of today's torpedo.

4. The torpedo boat *David* attacked the *New Ironsides* off Charleston, October 5, 1863.

5. Scheibert refers to the Billinghurst Requa Battery, "a 25-barrel volley gun." The barrels fired simultaneously and "could be moved laterally for 'spread'" (Coggins, *Arms and Equipment*, 44).

6. It remains unclear—perhaps a typographical error explains—why Scheibert says "10-pound Parrott," referring to the Swamp Angel, the huge 200-pounder, famous on the Union side, notorious on the Confederate.

7. Practice began as early as August 12, damaging the fort enough at times to seem like the real thing. Scheibert may be referring to the "first great bombardment," beginning (accordingly to Long) on August 17. See Long, *Civil War Day by Day*, 396–98.

8. A city and strategic site in the Franco-Prussian War, besieged by the Prussians for 108 days.

9. Scheibert refers to their use at the battles of the Wilderness and Spotsylvania and on the Petersburg campaigns, the last including the mine assault described by Scheibert below. Pioneering meant little in 1864 at Cold Harbor, in the valley, and on the Atlanta campaign.

10. The incident seems to have occurred during Stuart's first ride around McClellan, June 12–15, 1862. Von Borcke's "Memoirs," September 1865, 265. The original English has been supplied.

## Chapter 7. Medical Service

1. The systems began with an approximation of these distinctions. After 1862 the South established general hospitals and resembled the North more than Scheibert seems to recognize. Near the end of the war the systems again diverged

as collective medical care in the South, especially at the general hospitals, suffered with the rest of the Confederacy. See Horace Herndon Cunningham, *Doctors in Gray: The Confederate Medical Service*, 45–69. Richmond's Chimborazo hospital, for example, "then the largest in the world, treated in all some 76,000 patients" (Coggins, *Arms and Equipment*, 117).

2. Scheibert reports a better medical service than the typical description of North and South. Still, by his time (1863), McClellan's reorganization had improved Union medical service, especially on field and in removal from it. The Confederate service had developed similarly under Hunter Holmes McGuire and others. See Coggins, *Arms and Equipment*, 115–17; Cunningham, *Doctors in Gray*, 3–8, 114, 116–27, 247–66; George Worthington Adams, *Doctors in Blue: The Medical History of the Union Army*, 59–111.

3. The events occurred on May 3, 1863.

4. C. Bacon, "Second Extract from a Narrative of His Services in the Medical Staff," in U.S. Surgeon General's Office, *The Medical and Surgical History of the War of the Rebellion*, 137; Philip Adolphus, "Third Extract from a Narrative of His Services in the Medical Staff," in *Medical and Surgical History*, 138. Both Bacon's and Adolphus's words have been quoted in the English of the original. "No work of this character, of equal magnitude, has even been undertaken," let alone equaled, anywhere in the world (Joseph K. Barnes, Surgeon General, "Prefatory," *Medical and Surgical History*, vol. 1, pt. 1, p. viii).

5. Scheibert's volunteers may have been the civilians—ambulance drivers, nurses, surgeons—who worked for the armies, some as volunteers. See Coggins, *Arms and Equipment*, 116, 118; Adams, *Doctors in Blue*, 51–53, 158–160, 176–93; Cunningham, *Doctors in Gray*, 71–76.

McParlin's English from his original report has been given, and spelling, punctuation, and other peculiarities of style and mechanics retained. "Report, January 14 to July 31, 1864," dated November 28, 1864, in *Medical and Surgical History*, vol. 1, pt. 1, pp. 149–52.

6. *Medical and Surgical History*, vol. 1, pt. 1, 152.

## Chapter 8. Navies

1. Scheibert, sent to observe the war, could have chosen to stay in the North and observe the Union armies, or observed the armies of both sides. He chose to study the war from and with the South only; he loved the South as elitist and aristocratic but disliked the North as democratic and proletarian. Still, when writing this book, he kept the promise made at the start: "neither partial nor in anger" but "a dispassionate critique."

2. That is, Fort Calhoun, on a shoal three and a half miles offshore in Hampton Roads, therefore a distance from the action. Vessels there would be far from where needed. Fort Calhoun, later called Fort Wool, went by Rip Raps for the sound of waves on its shores. See William Chapman White and Ruth White, *Tin Can on a Shingle*, 8.

3. "Ten pieces of light ordnance" means two six-inch and two seven-inch muzzle-loading rifled cannon, and six nine-inch smoothbored guns. See Paul H. Silverstone, *Warships of the Civil War Navies*, 202.

4. Scheibert, seeming unsure of both numbers and kinds of vessels, at first refers to "seven frigates and steam-powered sloops," later to "four monitors," and never gives a total flotilla. The total may have been as many as eighteen of

various kinds. The USS *Tecumseh* hit the torpedo, prompting Farragut's "Damn the torpedoes, full speed ahead!" (Long, *Civil War Day by Day*, 551).

5. See Scheibert's study of it, below.

6. Scheibert continues the argument for wood over iron in warships and with reference to Germany and not our Civil War. He wants wooden ships built and equipped to modern standards. They will outdo iron, he says. He leaves no doubt as to his erudition and his skill in advocacy, but wood seemed done for when the *Monitor* and the *Merrimac* exchanged their first shots. Indeed, he admitted earlier in the chapter, "wood must yield to iron." The rest of his plea for wood has been deleted.

## Chapter 9. War All the Way: Mostly East

1. Missouri and Kentucky, though rebellious enough, did not secede; they were Border States.

2. Scheibert could more accurately have called Harpers Ferry an armory and Norfolk a naval base. Seizing Norfolk, the South got an arsenal and more, including a dry dock, an industrial plant, and vessels of the fleet. Long, *Civil War Day by Day*, 61, 63.

3. Scheibert's numbers seem wrong here. Lincoln's call for 75,000 would have brought this total to 90,000, probably too low a number.

4. McClellan, strictly speaking, created the new army around Washington, rather than in Virginia: around the Union capital rather than in the Confederacy.

5. Again Scheibert expresses his peculiar understanding of the Border States. Kentucky came closest to his description of the "neutral" states. See Edward Conrad Smith, *The Borderland in the Civil War*, 263–312.

6. Here, as in the rest of this paragraph and in many paragraphs to follow, Scheibert simplified what happened, thereby brought events sometimes into unusual proximity, and carried them beyond the year in question. His purpose directs him to compress and adjust: not to write a general chronicle, but to empasize the war's military side.

7. Or the Yorktown lines, passing through Williamsburg, drawn to hold McClellan south of the Chickahominy and away from Richmond.

8. If by "hostile government" Scheibert means Lincoln, his cabinet, and McClellan's superiors, their hostility produced mostly fighting among themselves, thereby perhaps less support for McClellan. In pages to come, Scheibert refers again to the Lincoln administration as harmful to the conduct of the war. Probably, to the elitist Scheibert, Lincoln, as a leader "of the people, by the people, for the people," could do nothing right, including lead a nation in wartime.

9. Probably the thirtieth. See Long, *Civil War Day by Day*, 234.

10. In point of fact, Stuart did not go the length of Maryland but swept most of the western part.

11. Contrary to the way Scheibert and others refer to the nickname, it did not originate as a term of affection or as a description of a pugnacious attitude; however, it may have proved accurate for aggressive leadership later. The name seems to have begun as a journalistic accident. It brought from Hooker the "plea to the press": "Don't call me Fighting Joe, for that name has done and is doing me incalculable injury," causing "a portion of the public to think that I am a hot headed, furious young fellow, accustomed to making furious and needless

dashes at the enemy" (Walter H. Hebert, *Fighting Joe Hooker*, 91, see also pp. 90, 318–19 n. 76).

12. That is, abandon the heights at Fredericksburg and move to protect the capital from Hooker's advance. Burnside had failed against the Confederates on the heights. Since then the Confederates had fortified them even more, earning for them the title *redoubt*.

13. That is, the Confederates had been on the move, cavalry screening them. Union cavalry arrived to investigate. The mounted forces fought "the greatest cavalry battle on American soil" (Long, *Civil War Day by Day*, 363. Compare Freeman, *Lee*, 3:32). Scheibert with Von Borcke wrote about the battle in *Die grosse Reiterschlacht*.

14. This brief, elliptical paragraph refers to the Bristoe (Confederate) and Mine Run (Union) campaigns, October and November 1863.

15. Unlike Jackson, Longstreet recovered from wounds and returned to command.

16. In point of fact, May 19. Perhaps Scheibert wrote June 19 because he was confused by the jumbled surge of events at this stage of a complicated situation.

17. Scheibert remains a month off; the date should be June 3, 1864. Long calls that battle at Cold Harbor "one of the bloodiest battles of history" (*Civil War Day by Day*, 513).

18. That is, he continued the siege at Petersburg as the way to Richmond.

## Chapter 10. War in the West: Combined Operations

This chapter consists mostly of what I have called Study Two, his separate work on combined operations. I have used the modern term *combined operations* for Scheibert's (literally translated) "army-navy cooperation." The first section comprises two short chapters of Scheibert's, edited and rearranged to eliminate most of his references to Europe and to emphasize combined operations on the Mississippi. He shows his credentials and states his purpose: "Having studied the sources and referred to the maps in abundance, I could write the history of the lower Mississippi as a theater of the American Civil War. I limit myself to what we need to understand army-navy cooperation there, the essentials, nothing else. Those and only those details remain that could not be omitted in making or clarifying the point, none other than serve the purpose. Thus you have below in the conciset terms my review and analysis of combined operations in that grand, eventful, consequential struggle for the American heartland. Please because of the elliptical brevity refer to maps as needed."

1. That is, probably, Admiral Andrew H. Foote in the capture of Fort Henry on the Tennessee in February 1862 and Admiral David D. Porter at Arkansas Post in January 1863.

2. Instead of Nevada, Scheibert ought to have said the Rockies and their ranges.

3. "See the appendix on them" (Scheibert).

4. High-angle fire—probably the kind Francis Scott Key called "bombs bursting in air"—out of mortars, smoothbored guns, and muzzle-loading rifles. See John D. Milligan, *Gunboats down the Mississippi*, 39; Silverstone, *Warships*, 158–59.

5. "General Manning F. Force, *From Fort Henry to Corinth*, says 24,400. That little book, confused and unreliable, must be used with caution, however. See

Francis V. Greene, *The Mississippi*, in the same series: well written and reliable" (Scheibert). He refers to the two as part of the Scribner's series *Campaigns of the Civil War*, volumes 2 and 8, respectively.

6. Scheibert implies an exchange of messages and an agreement to cooperate. By the records available to him, however, Grant (December 13, 1861) wrote Foote of the probability of attack and "would respectfully ask your cooperation with gunboats" (U.S. War Department, *Official Records . . . Navies*, ser. 1, vol. 22, pp. 461–62). Foote's telegram of the next day, to Captain George D. Wise— "Inform General Cullum that the gunboats and twenty of the mortar boats are here ready for examination" (p. 462)—could be read that Foote would cooperate as Grant asked. The telegram refers, however, to an ongoing concern between Foote and Callum, not Grant's request. Messages of January and February 1862 show Grant and Foote communicating about cooperation (esp. pp. 482, 489–90). They cooperated, though Foote participated less than he would have preferred.

7. The attackers may have numbered twice the defenders. The Confederates may have surrendered because they were outnumbered two to one, surrounded, and confronted by Grant's "no terms except unconditional surrender." Still it is not clear why Floyd turned back after the soldiers may have found a way out. He may have been attacked and driven back and not wanted to return. Yet, as Scheibert suggests, if so many could escape, "some simply walked away," why did so many stay and surrender? See Herman Hattaway and Archer Jones, *How the North Won: A Military History of the Civil War*, 69–75; Richard E. Beringer et al., *Why the South Lost the Civil War*, 123–24. "Some simply walked away" is in Long, *Civil War Day by Day*, s.v., "February 16," p. 172.

8. See Gray to Colonel Lewis G. De Russy, August 20, 1861. U.S. War Department, *Official Records*, ser. 1, vol. 1, pp. 390–91.

9. Compare "I passed up the river this morning. . . . I am satisfied that it is not possible for us to take Vicksburg without an army force of twelve or fifteen thousand men" (Farragut, Report to Secretary Welles, U.S. War Department, *Official Records . . . Navies*, ser. 1, vol. 18, p. 588 [June 28, 1862]).

10. Scheibert always refers to General Joseph E. Johnston as Joe Johnston. Another sign of his familiarity, friendship, even intimacy with Confederate generals?

11. General Trenton remains unidentified and may not have existed, his name perhaps a confusion with Trenton, Tennessee. Compare Long, who says Forrest got the worse of the scuffle but "managed to escape" (*Civil War Day by Day*, 303).

12. Scheibert has "A. W. Morgan" but must have meant George W. Morgan. See, for example, Hattaway and Jones, *How the North Won*, 313.

13. Here begins Scheibert's discussion of the Battle of Chickasaw Bayou.

14. Johnson's plantation: Scheibert calls it Johnson's House, but see Milligan, *Gunboats*, 107.

15. Johnston commanded the Department of the West—Pemberton, under him, the Department of Mississippi, Tennessee, and East Louisiana.

16. That is, the *Queen* hit the *Vicksburg*, caused some damage and started fires on her, but the *Vicksburg* stayed on line.

17. Scheibert, in an obvious mistake or misprint in the original, has 70,000 as the total for these forces.

18. Scheibert gives it in full, translated into German. This is the original, from U.S. War Department, *Official Records . . . Navies*, ser. 1, vol. 24, pp. 607–9. Note that it is labeled not "letter," but "general order":

General order of Acting Rear-Admiral Porter, U.S. Navy, preparatory to the attack.

> MISSISSIPPI SQUADRON,
> *Flagship Benton, Ashwood Landing*
> *Mississippi River, April 27, 1863.*

In going into action with the forts at Grand Gulf the following orders will be observed:

It is reported that there are four positions where guns are placed, in which case it is desirable that all four places should be engaged at the same time. The *Louisville, Carondelet, Mound City,* and *Pittsburg* will proceed in advance, going down slowly, firing their bow guns at the guns in the first battery on the bluff, passing 100 yards from it, and 150 yards apart from each. As they pass the battery on the bluff they will fire grape, canister, and shrapnel, [set to fire] cut at one-half second, and percussion shell from rifled guns.

The leading vessel (*Louisville*) will round to at the next battery, keeping the bow presented, as if carried past by the current, come up again and engage it. The next vessel will engage the third battery, and the next the fourth, the last vessel preparing to double on what appears to be the heaviest of the lower batteries. The *Benton* and *Tuscumbia* will attack the upper batteries on the bluff going down slowly and firing shell with 5-second fuzes, one bow gun to be loaded and fired with canister. The *Lafayette* will drop down at the same time, stern foremost, until within 600 yards, firing her rifled guns with percussion shells at the upper battery. The *Tuscumbia* will round to outside the *Benton*, not firing over her while so doing; after rounding to, she will keep astern and inside of the *Benton*, using her bow guns while the *Benton* fires her broadside guns. The *Tuscumbia and Benton* [sic] will also fire their stern guns at the forts below them whenever they will bear, using shell altogether.

The *Louisville, Carondelet, Mound City,* and *Pittsburg* will also keep their stern guns trained sharp on the batteries on their quarter, firing deliberately and trying to dismount the enemy's batteries.

The four vessels leading will take position 100 yards from the beach or landing, firing 5-second shell, and one gun firing altogether with shrapnel cut at 2–1/2 seconds, unless the commanders see that it is too long or too short.

If I find that the upper battery is soon silenced I will hoist the guard flag and blow a long whistle for the *Lafayette* to drop down and assist the four steamers at the lower batteries. I will blow long and continued whistles, without any flag, until the order is obeyed. The *Lafayette* and *Tuscumbia* must concentrate their fire on what appears to be the heaviest battery below, and obtain such a distance as will enable them to fire accurately at the guns of the enemy.

The stern, side, and bow guns must be used by all the vessels when practicable and when there is no possibility of firing into each other.

If a battery is silenced perfectly, each vessel must pick out the next one the commander thinks troublesome and get a position where she will not interfere with anyone else; when it is practicable, form a line abreast and bring all the bow guns to bear on one place; that will soonest end the fight.

Every vessel should be well packed with hammocks, bags, and awnings around the pitmans. Every precaution should be taken against the houses on deck taking fire; water buckets and tubs should be kept filled all about the spar deck; an officer or trusty person to look out and report if fire breaks out.

The guns must be run into a taut breeching and the men cautioned about sticking themselves out of the ports when loading. Coolness in firing is recommended; let not a shot be thrown away.

If any vessel is disabled in machinery, let her drop her anchor at once and fight as long as she can; one anchor to be kept ready for this purpose.

If sharpshooters appear on the hillside or in rifle pits, throw shrapnel cut to the proper length at them, calculating that the initial velocity is 1,200 feet per second.

The *Benton* will take position as circumstances may require. Commanders will send for their pilots and explain to them what is to be done, also explain the position of the forts to the officers, and let every man know what he has to do. Let the officers explain the same thing to the crew at quarters.

The water is falling; no vessel will anchor for the present in less than 4 fathoms water, and the lead must be frequently used.

Very respectfully,
    DAVID D. PORTER,
    *Acting Rear-Admiral, Commanding Mississippi Squadron.*
  Captain HENRY WALKE,
    *Commanding Lafayette.*

19. In fact the Confederacy eliminated both from this campaign but destroyed neither. Both served out the war. See Silverstone, *Warships,* 70–71, 161.

20. Scheibert does not identify the author.

21. Achille Francois Bazain and Marie Edme Patrice de MacMahon lost to the Prussians in 1870 while defending the sites named. Osman Nuri Pasha, a Turkish general, surrendered to the Russians at Plevin in the Russo-Turkish War of 1877–1878.

## Epilogue: The Lesson of the War

1. This copy of Lee's order is not a translation of Scheibert's translation; it comes from Frank Moore, *The Rebellion Record: A Diary of American Events,* vol. 7, doc. 82, p. 323.

Pro-Southern passion tempts Scheibert to exaggerate more in this than any other chapter. If he cannot win the war for the South, he will save face for it in the tribunal of history; he will be the character witness for the convicted awaiting sentence. If the South lost the war, to him it won more battles. If it lost the war, it had better people and a nobler cause. So what if it lost on the battlefield; might does not make right.

2. Johann Tserklaes, Count of Tilly, a general in the Thirty Years War, sacked Magdeburg in 1631. In 1632, Gustavus Adolphus smashed Tilly's forces in two battles. Tilly died of wounds suffered in the second. Few instances in the history of war resound louder than Napoleon's invasion of Russia with 422,000 men, taking Moscow with the 100,000 who made it that far, then retreating with something less than the 100,000, to leave Russia's ice and snow with only 10,000.

3. The reference seems likeliest to John McCausland's capture with cavalry and burning of Chambersburg, Pennsylvania, July 30-31, 1864, after demanding and being refused $500,000 in paper or $100,000 in gold as "reparation for [David] Hunter's depredations in the Shenandoah" earlier in July and in June 1864. McCausland, in turn, took a beating on August 1, 2, and 3, 1864. See Long, *Civil War Day by Day*, 548, 549, 550 (quoted phrase, 548).

4. In 1862, Major General Benjamin F. Butler took over New Orleans in May, ruled until December, and from citizens prompted a phrase to label what he had done: "bestial acts." In 1864, Major General David Hunter's destructive raids in the Shenandoah included the burning of the Virginia Military Institute, June 11. In 1863, Major General Robert H. Milroy committed similar acts. Major General Louis Blenker's division, transferred from the Army of the Potomac to Frémont's Mountain Department in West Virginia, raided and looted as they went, March-April 1862. In 1864 and early 1865, Major General Philip H. Sheridan led Union cavalry on destructive raids in the Shenandoah. David P. Conyngham, *Sherman's March through the South*, 268-69. Conyngham's language has been supplied, not a translation of Scheibert's translation.

5. That is, Major (Brevet) George Ward Nichols (1837-1885), *The Story of the Great March*.

6. U.S. Congress, *Joint Committee Report*, 1:iii-iv (May 22, 1865).

7. Henry Charles Fletcher, *History of the American War*, 2:130, 227-29, 289:

Pope decided on retreating to the Rappahannock. Then were found the ill effects to his own army of his previous order [Referring to the orders of July 18—"General order, No. 5. Hereafter, as far as practicable, the troops of this command will subsist on the country in which their operations are carried on, &c., &c." This order, and others which related to the treatment of the inhabitants, resulted in plundering, although such was not the probable intention of General Pope.] which virtually had almost sanctioned plundering and depredation. The troops had become demoralised, the country on the line of march had been devastated, and the country people embittered to the highest degree against the army at whose hands they had received such injuries. General Pope was forced for the well-being of his own forces to explain more fully his former order, and to take measures to prevent depredations on the property of the inhabitants.

. . . . . . . . . . . . . . . . . . . . . .

At New Orleans the reign of General Butler was drawing to a close; his tyranny had not only provoked the most deadly hatred in the South, but had even kindled a feeling of shame in the hearts of many of the more respectable of the Northern people, whilst his illegal acts in regard to a foreign nation had called forth a strong remonstrance from the French Government. His removal was determined on, and General Banks, who although he had shown little military capacity, was yet esteemed as a gentleman and an able administrator, was appointed to succeed him.

In his farewell to the people of New Orleans, General Butler sought justification for his own acts of tyranny, which he termed merited severity, by a category of charges against European nations, ransacking their history for a period of two centuries to find instances of cruelty which he could contrast with his own mild exercise of authority. His proclamation contains the following paragraph, which may be read as a specimen of the writings of some of the civilian generals of America:—"I do not feel that I have erred in too much harshness, for that harshness has ever been exhibited to disloyal enemies of my country, and not to loyal friends. To be sure I might have regaled you with the amenities of British civilisation, and yet been within the supposed rules of civilised warfare. You might have been smoked to death in caverns, as were the Covenanters of Scotland, by the command of a general of the royal household of England; or roasted like the inhabitants of Algiers during the French campaign. Your wives and daughters might have been given over to the ravisher, as were the unfortunate dames of Spain in the Peninsular War; or you might have been scalped and tomahawked, as our mothers were at Wyoming by the savage allies of Great Britain in our own revolution; your property could have been turned to indiscriminate 'loot,' like the palace of the Emperor of China; works of art which adorned your buildings might have been sent away like the paintings at the Vatican; your sons might have been blown from the mouths of cannon like the sepoys of Delhi; and yet all this would have been within the rules of civilised warfare, as practised by the most polished and the most hypocritical nations of Europe. For such acts the records of the doings of some of the inhabitants of your city towards the friends of the Union, before my coming, were a sufficient provocation and justification. But I have not so conducted [sic]." General Butler then enumerates the mild punishments he was forced to inflict, and also the great benefits he had conferred on the city of New Orleans—benefits which were certainly not appreciated by its inhabitants.

. . . . . . . . . . . . . . . . . . . . . . . . . . . . .

During this time Sheridan was laying waste the country, converting the most fruitful district of Virginia into a complete wilderness,—not only devastating the crops and carrying off the cattle, but burning the barns, mills, and agricultural implements, thus equalling, if not surpassing, in deeds of rapine and violence, the commanders whose names have been held up to reprobation for similar acts in the ways of past centuries. What Hunter had spared, Sheridan destroyed; and not content with inflicting temporary ruin on the country and its inhabitants, did what lay in his power to prevent a return to prosperity when peace should again visit the land, and the population reseek their abandoned houses. The following extract is taken from General Sheridan's dispatch:—"I have destroyed over 2,000 barns filled with wheat, hay, and farming implements; over 70 mills filled with flour and wheat; have driven in front of the army over 4,000 head of stock, and have killed and issued to the troops not less than 3,000 sheep." These acts of General Sheridan, Mr. Swinton, in his *History of the Army of the Potomac*, justly reprobates, comparing them to the deeds of the French in the Palatinate [in the Napoleonic Wars].

8. Probably the *Tagebüchern und Aufzeichnungen* of Heinrich von Brandt (1789–1868), Prussian general in the Napoleonic era, edited by his son, Heinrich von Brandt, 3 vols. (Berlin: Mittler, 1870–1882). Scheibert's "recent campaigns over here" refers to the Prusso-Danish, Austro-Prussian, and Franco-Prussian wars of 1864, 1866, and 1870–1871, respectively.

# Appendix

1. Scheibert may have confused the two boats here. Earlier he calls the *Venus* a transport and suggests the *Undine* as the important one. The *Undine*, 179 tons (Silverstone, *Warships*, 179), could be called one of the biggest in the Mississippi River service. Silverstone says nothing about the *Venus*.

2. Scheibert has 133,000, probably a mistake for 33,000, especially as his total is 39,300, high as such figures usually run, but nearer reality than 139,300. But, his addition being so grossly wrong, the whole table must be a tissue of errors. I have corrected minor ones in the tables shown. They, too, cannot be trusted therefore. Scheibert's values lie elsewhere than in his generally and at best inexact statistics.

3. Scheibert says *mit senkrechten Bollwerken* (with perpendicular bulwarks), an accurate description of how they looked. Another American term, used with such boats, is *timberclad,* because armored in wood. See Silverstone, *Warships,* 158–59, for specifics on this trio.

4. In point of fact the *Merrimac* had been a warship from the start, designed by Lenthall, built in Boston (1854), and considered with her sisters "to be superior to any warship in the world" (ibid., 27). She had been burned to the waterline to prevent capture at Norfolk. The Confederates rebuilt her as the *Virginia* that fought the *Monitor.* If her origins contradict Scheibert, the rebuilding and especially the armoring did serve as he says: a model for refitting civilian craft. In addition, notice that these converted merchant vessels, ironclads, should not be confused with the three that Scheibert describes as timberclads, *Conestoga, Lexington,* and *Tyler,* armored chiefly with wood. See Silverstone, *Warships,* 147, 158–59.

5. Ibid., 151–53.

6. Ibid., 161–62.

7. Ibid., 158–59.

8. Walkes's original has not been identified. This is a paraphrase, in translation, of Scheibert's quotation in German.

# BIBLIOGRAPHY

Consulted for the editing and translating of this work
and
Recommended for the reading of it

## Bibliographies and Maps

Coulter, E. Merton. *Travels in the Confederate States: A Bibliography.* Norman: University of Oklahoma Press, 1948.
Dornbusch, Charles E., comp. *Military Bibliography of the Civil War.* 3 vols. New York: New York Public Library, 1967–1972.
Eicher, David J. *The Civil War in Books: An Analytical Bibliography.* Urbana: University of Illinois Press, 1997.
Freeman, Douglas Southall. *The South to Posterity: The Writings of Confederate History.* New York: Scribner's, 1939.
McPherson, James M. *The Atlas of the Civil War.* New York: Macmillan, 1994.
Nevins, Allan, James I. Robertson, Jr., and Bell I. Wiley, eds. *Civil War Books: A Critical Bibliography.* 2 vols. Baton Rouge: Louisiana State University Press, 1967–1969.
U.S. Military Academy, Department of Military Art and Engineering, comp. Brigadier General Vincent J. Esposito, ed. *The West Point Atlas of American Wars.* Vol. 1, s. v. *Civil War.* New York: Henry Holt, 1995.

## Works by Justus Scheibert

**Books**

*Einfluss der neuesten Taktik und der gezogenen Waffen auf den Festungskrieg* (The Influence of the Latest Tactics and of Rifled Arms on Fortification in Warfare). Berlin: Mittler, 1861.
*La Guerre civil aux États-Unis d'Amérique: Guerre de la sécession.* Publication de la Reunion des Officiers. Translated by J. Bornecque. Paris: Dumaine, 1876. Translation of *Der Bürgerkrieg in den nordamerikanischen Staaten: Militärisch beleuchtet für den deutschen*

*Offizier* (The Civil War in America: A Military Handbook for the German Officer). Berlin: Mittler, 1874.

*Die Befestungskrieg und die Lehre von Kampfe* (Fortification in Warfare and the Theory of Military Science). 3 vols. Berlin: Mittler, 1880–1886.

Borcke, Heros von, and Justus Scheibert. *Die grosse Reiterschlacht bei Brandy Station, 9 June 1863* (The Big Cavalry Battle at Brandy Station, June 9, 1863). Berlin: Kittel, 1893.

*Der Krieg Deutschland und Frankreich . . . 1870/71*. Leipzig: Fock, 1888. Translated as *The Franco-German War, 1870–71* by Major J. A. Ferrier and Mrs. Ferrier. Chatham: Royal Engineers Institute, 1894.

*Illustriertes deutsches Militär-Lexikon* (Illustrated Dictionary of German Military Science). Edited by Justus Scheibert et al. Berlin: Pauli's, 1897.

*Kaiser Wilhelm I. und seine Zeit* (Emperor William I and His Times). 2 vols. Berlin: Becker, 1898.

*Der Freiheitskampf der Buren und die Geschichte ihres Landes* (The Boer War, with a History of the Boers' Homeland). 2 vols. Berlin: Schroeder, 1900.

*Der Krieg in China, 1900–1901; nebst einer Beschreibung der Sitten, Gebräuche und Geschichte des Landes* (The War in China [the Boxer Rebellion], with a Description of Chinese Customs, Habits, and History). Berlin: Schroeder, 1901–1902.

*Mit Schwert und Feder: Erinnerungen aus meinem Leben* (With Sword and Pen: Memoirs). Berlin: Mittler, 1902.

*Die Kriege von 1864 und 1866* (The Wars of 1864 and 1866 [the Prusso-Danish and Austro-Prussian Wars]). Berlin: Weller, 1904.

*Seven Months in the Rebel States during the North American War, 1863.* (*Sieben Monate in den Rebellion-Staaten während des nordamerikanischen Krieges 1863.*) Edited by William Stanley Hoole. Translated by Joseph C. Hayes. Tuscaloosa: Confederate Publishing Company, 1958. Stettin: Nahmer, 1868.

**Articles**

*Jahrbücher für die deutsche Armee und Marine 1–47* (1871–1918), then *Monatshefte für die deutsche Wehrmacht* (January and February 1919), ceased publication as *Monatshefte für Politik und Wehrmacht* (1922): includes about fifty articles, lectures, speeches, and reviews, some reprinted in later issues, on the American Civil War.

**In the *Southern Historical Society Papers (SHSP):***
"To the Rev. Williams Jones, D. D., Secretary, Southern Historical Society," vol. 2 (December 1876): 318.

"Letter from Major Scheibert, of the Prussian Royal Engineers," vol. 5 (January–February 1877): 90–91.

"Review of *Allen's History of the Valley Campaign*," vol. 11 (July 1883): 327–28.

"The Work of the Southern Historical Society in Europe: Letter from Major Scheibert," vol. 9 (October-November-December 1881): 571.

## General

**Books**

Acton, Lord Harold. "The Civil War in America: Its Place in History." In *Acton's Historical Essays and Studies*. London: Macmillan, 1907. Reprint. Freeport, N.Y.: Books for Libraries, 1967.

Adams, George Worthington. *Doctors in Blue: The Medical History of the Union Army in the Civil War*. New York: Schuman, 1952.

*The American Heritage Picture History of the Civil War*. New York: American Heritage, 1960.

Bartlett, Merrill L. *Assault from the Sea: Essays in the History of Amphibious Warfare*. Annapolis: Naval Institute Press, 1983.

Batchelor, John, and Ian V. Hobb. *Artillery*. New York: Scribner's, 1972.

*Battles and Leaders of the Civil War*. Edited by Clarence C. Buel and Robert U. Johnson. 4 vols. New York: Century, 1884–1887. Reprint. 4 vols. New York: Barnes, 1957.

Bellet, Paul Pecquet du. *The Diplomacy of the Confederate Cabinet at Richmond and Its Agents Abroad*. Tuscaloosa: Confederate Publishing Company, 1963.

Beringer, Richard E., et al. *Why the South Lost the Civil War*. Athens: University of Georgia Press, 1986.

Birkheimer, William E. *Historical Sketch of . . . the Artillery, United States Army*. N.p.: Chapman, 1884. Reprint. West Point Military Library. New York: Greenwood, 1968.

Black, Robert C. III. *The Railroads of the Confederacy*. Chapel Hill: University of North Carolina Press, 1952.

Blackford, W[illiam]. W. *War Years with Jeb Stuart*. New York: Scribner's, [1945?].

Boatner, Mark Mayo. *The Civil War Dictionary.* New York: David McKay, 1959.

Borcke, Heros von. "Memoirs of the Confederate War for Independence." 10 installments. *Blackwood's Edinburgh Magazine* 98 and 99 (September, October, November, December 1865, and January, February, March, April, May, June 1866): pp. (respectively) 269–88, 389–437, 557–80, 635–55, 83–102, 173–96, 307–21, 448–68, 543–64, 747–70. Published under same title, 2 vols. Edinburgh: Blackwoods, 1866.

Borcke, Heros von, and Justus Scheibert. *Die grosse Reiterschlacht bei Brandy Station.* Berlin: Kittel, 1893.

Bowman, John S. *The Civil War Day by Day.* N.p.: Dorset, 1989.

Brackett, Albert G. *History of the United States Cavalry.* New York: Harper, 1865. Reprint. West Point Military Library. New York: Greenwood, 1968.

Bradlee, Francis B. C. *Blockade Running during the Civil War: The Effect of Land and Water Transportation on the Confederacy.* Salem, Mass.: Essex Institute, 1925.

Brinton, John H. *Personal Memoirs of John H. Brinton, Major and Surgeon, U.S.V., 1861–1865.* New York: Neale, 1914.

Brooks, Stewart. *Civil War Medicine.* Springfield, Ill.: Thomas, 1966.

Buehr, Walter. *Firearms.* New York: Crowell, 1967.

Bull, Stephen, *An Historical Guide to Arms and Armor.* Edited by Tony North. New York: Facts on File, 1991.

Burton, E. Milby. *The Siege of Charleston, 1861–1865.* Columbia: University of South Carolina Press, 1970.

Carter, Samuel. *The Last Cavaliers: Confederate and Union Cavalry.* New York: St. Martin's, 1979.

Casey, Silas. *Infantry Tactics for the Instruction, Exercise, and Manoeuvers of the Soldier, a Company, Line of Skirmishers, Battalion, Brigade, or Corps d'Armee.* 3 vols. New York: Van Nostrand, 1863.

Chambers, Lenoir. *Stonewall Jackson.* 2 vols. New York: Morrow, 1959.

Chesney, Charles Cornwallis. *A Military View of Recent Campaigns in Virginia and Maryland.* 2d ed. 2 vols. London: Smith, Elder, 1863–1865.

Coggins, Jack. *Arms and Equipment of the Civil War.* Garden City, N.Y.: Doubleday, 1962.

Connelley, Thomas L. *The Marble Man: Robert E. Lee and His Image in American Society.* New York: Knopf, 1977.

Conyngham, David P. *Sherman's March through the South.* New York: Sheldon, 1865.

Cornish, Dudley. *The Sable Arm: Negro Troops in the Union Army.* New York: Longmans, Green, 1956.

Corsan, W. C. [An English Merchant, pseud.]. *Two Months in the Confederate States: Including a Visit to New Orleans.* London: Bentley, 1863.

Cunningham, Horace Herndon. *Doctors in Gray: The Confederate Medical Service.* Baton Rouge: Louisiana State University Press, 1958.

Davis, Burke. *Jeb Stuart: The Last Cavalier.* New York: Rinehart, 1957.

Davis, William C. *The Embattled Confederacy.* New York: Doubleday, 1982.

DeLeon, Thomas Cooper. *Fours Years in Rebel Capitals: Life in the Southern Confederacy.* Mobile: Gossip Printing Company, 1890.

Derry, Joseph Tyrone. *Story of the Confederate States.* Richmond: Johnson, 1895.

Diagram Group. *Weapons: An International Encyclopedia.* Edited by David Harding. New York: St. Martin's, 1980.

Donald, David, ed. *Divided We Fought: A Pictorial History of the Civil War.* New York: Macmillan, 1956.

Dupuy, Trevor N. et al. *Dictionary of Military Terms: The Language of Warfare and Military Institutions.* New York: Wilson, 1986.

Dupuy, Trevor N., and Ernst R. Dupuy, eds. *The Encyclopedia of Military History.* 2d ed. New York: Harper, 1985

Estván, Bela. *War Pictures from the South.* 2 vols. New York: Appleton, 1863.

Evans, Charles A., ed. *Confederate Military History.* Atlanta: n.p., 1899.

Faust, Patricia et al., eds. *Historical Times Illustrated Encyclopedia of the Civil War.* New York: Harper & Row, 1986.

Ferris, Norman B. *Desperate Diplomacy.* Knoxville: University of Tennessee Press, 1976.

Fite, Emerson D. *Social and Industrial Conditions in the North during the Civil War.* 1910. Reprint. AMS Press, 1982.

Fletcher, Henry Charles. *History of the American War.* 3 vols. London: Bentley, 1865.

Foote, Shelby. *The Civil War: A Narrative.* 3 vols. New York, Random House, 1958–1974.

Force, Manning F. *From Fort Henry to Corinth.* Vol. 2 of *Campaigns of the Civil War.* New York: Scribner's, 1881.

Fowler, William L., Jr. *Under Two Flags: The American Navy in the Civil War.* New York: Norton, 1990.

Freeman, Douglas Southall. *R. E. Lee: A Biography.* 4 vols. New York: Scribner's, 1934–1935.

Fuller, J[ohn] F. C. *The Generalship of Ulysses S. Grant.* 2d ed. Bloomington: Indiana University Press, 1958.

———. *Grant and Lee: A Study in Personality and Generalship.* Bloomington: Indiana University Press, 1957.

———. *A Military History of the Western World.* 3 vols. New York: Funk & Wagnalls, 1954–1956.

Geyl, Pieter. "The American Civil War and the Problem of Inevitability." In *Debates with Historians.* Groningen, Netherlands: Wolters, 1955.

Gibbon, John. *The Artillerist's Manual.* New York: Van Nostrand, 1860. Reprint. West Point Military Library. Westport, Conn.: Greenwood Press, 1971.

Gillmore, Q. A. *Engineer and Artillery Operations against the Defences of Charleston Harbor in 1863.* New York: Van Nostrand, 1868.

Girard, Charles. *A Visit to the Confederate States of America in 1863: Memoir to His Majesty Napoleon III.* Tuscaloosa: Confederate Publishing Company, 1962.

Gosnell, H. Allen. *Guns on the Western Waters: The Story of River Gunboats in the Civil War.* Baton Rouge: Louisiana State University Press, 1949.

Greene, Francis V. *The Mississippi.* Vol. 8 of *Campaigns of the Civil War.* New York: Scribner's, 1882.

Hardee, William Joseph. *Rifle and Light Infantry Tactics for the Exercise and Maneuvers of Troops. . . .* 2 vols. Philadelphia: Lippincott, 1861.

Harris, William C. *Leroy Pope Walker: Confederate Secretary of War.* Tuscaloosa: Confederate Publishing Company, 1961.

Hattaway, Herman, and Archer Jones. *How the North Won: A Military History of the Civil War.* Urbana: University of Illinois Press, 1983.

Hebert, Walter H. *Fighting Joe Hooker.* Indianapolis: Bobbs-Merrill, 1944.

Henderson, G[eorge] F. R. *The Civil War: A Soldier's View.* Edited by Jay Luvaas. Chicago: University of Chicago Press, 1958.

———. *The Science of War: Essays and Lectures, 1891–1903.* Edited by Captain Neill Malcolm. New York: Longmans, Green, 1908.

———. *Stonewall Jackson and the American Civil War.* New York: Grosset & Dunlap, n.d.

Hoole, William S. Introduction to *Seven Months in the Rebel States*, by Justus Scheibert, pp. 7–16. Tuscaloosa: Confederate Publishing Company, 1958.

Hoole, W. Stanley. *Vizetelly Covers the Confederacy*. Tuscaloosa: Confederate Publishing Company, 1957.

Hughes, Nathaniel Cheairs, Jr. *General William J. Hardee*. Baton Rouge: Louisiana State University Press, 1965.

Jessup, John E., ed. *Encyclopedia of the American Military: History, Traditions, Policies, Institutions, and Roles of the Armed Forces*. 3 vols. New York: Scribner's, 1994.

Joinville, Francois [August Trognan, pseud.]. *History of the Civil War in America*. 4 vols. Philadelphia: Lippincott, 1875–1888. Translation of *Guerre d'Amérique*. 8 vols. Paris: Dumaine, 1874–1883.

Jones, Archer. *Confederate Strategy from Shiloh to Vicksburg*. Baton Rouge: Louisiana State University Press, 1961.

Jones, James P. *Yankee Blitzkrieg: Wilson's Raid through Alabama and Georgia*. Athens: University of Georgia Press, 1987.

Jordan, Donaldson, and Edwin J. Pratt. *Europe and the American Civil War*. Boston: Houghton Mifflin, 1931. Reprint. New York: Octagon Books, 1969.

Ketchum, Hiram. *General McClellan's Peninsula Campaign: Review of the Report of the Committee on the Conduct of the War*. New York: Journal of Commerce, 1864.

Kirke, Edmund [James R. Gilmore, pseud.]. *Among the Pines: or, The South in Secession*. New York: Gilmore, 1862.

———. *Life in Dixie's Land*. New York: Carleton, 1863.

La Bree, Benjamin, ed. *Campfires of the Confederacy: Anecdotes. . . .* Louisville: Courier-Journal, 1899.

———. *The Confederate Soldier in the Civil War, 1861–1865*. Louisville: Courier-Journal, 1895.

———. *The Lost Cause: A Confederate War Record*. 10 vols. Louisville: Courier-Journal, 1898–1904.

———. *Pictorial Battles of the Civil War . . . Upwards of One Thousand Engravings*. 2 vols. New York: Sherman, 1885.

LaBree, Benjamin, and Wright Marcus, eds. *Official and Illustrated War Record . . . Nearly One Thousand Pictorial Sketches*. Washington: Government Printing Office, 1899.

Lawrence, Rev. G[eorge] G. *Three Months in America in the Summer of 1863*. London: Whittaker, 1864.

Lincoln, Abraham. *Collected Works.* Edited by Roy P. Basler. 9 vols. New Brunswick, N.J.: Rutgers University Press, 1953.

Livermore, Thomas L. *Numbers and Losses in the Civil War.* Boston: n.p., 1900.

Long, E. B., and Barbara Long. *The Civil War Day by Day: An Almanac 1861–1865.* Garden City, N.Y.: Doubleday, 1971.

Longacre, Edward G. *Mounted Raids of the Civil War.* New York: Barnes, 1975.

Lonn, Ella. *Foreigners in the Confederacy.* Chapel Hill: University of North Carolina Press, 1940.

Lord, Francis A. *Civil War Collector's Encyclopedia.* Harrisburg, Penn.: Stackpole, 1963.

Luraghi, Raimondo. *A History of the Confederate Navy.* Translated by Paolo E. Coletta. Annapolis: Naval Institute Press, 1996.

Luvaas, Jay. *The Military Legacy of the Civil War: The European Inheritance.* Chicago: University of Chicago Press, 1959.

McFeely, William S. *Grant: A Biography.* New York: Norton, 1981.

Mackenzie, Robert. *America and Her Army.* New York: Nelson, 1865.

McPherson, James M. *The Negro's Civil War.* New York: Pantheon, 1965.

McWhiney, Grady. *Braxton Bragg and Confederate Defeat.* New York: Columbia University Press, 1969.

———. *Grant, Lee, Lincoln and the Radicals.* Evanston, Ill.: Northwestern University Press, 1964.

McWhiney, Grady, and Perry D. Jamieson. *Attack and Die: Civil War Tactics and the Southern Military Heritage.* University, Ala.: University of Alabama Press, 1982.

Meerheimb. F. von. *Shermans Feldzug in Georgien.* Berlin: Mittler, 1869.

Milligan, John D. *Gunboats down the Mississippi.* Annapolis: Naval Institute Press, 1965.

Mitchell, Donald W. *History of the American Navy.* New York: Knopf, 1946.

Mitchell, Joseph Brady, and Sir Edward S. Creasy. "Vicksburg." In *Twenty Decisive Battles of the World.* New York: Macmillan, 1964.

Moore, Frank. *The Rebellion Record: A Diary of American Events.* 11 vols. New York: Putnam, 1861–1863; Van Nostrand, 1864–1868.

Musicant, Ivan. *Divided Waters: The Naval History of the Civil War.* New York: HarperCollins, 1995.

Nagel, Paul C. *The Lees of Virginia: Seven Generations of an American Family*. New York: Oxford University Press, 1990.
Nevins, Allan. *The Statesmanship of the Civil War*. New York: Macmillan, 1953.
Nichols, George Ward. *The Story of the Great March*. New York: Harper, 1865.
Nichols, James L. *Confederate Engineers*. Tuscaloosa: Confederate Publishing Company, 1957.
Owsley, Frank L. *King Cotton Diplomacy: Foreign Relations of the Confederate States of America*. 2d ed. Chicago: University of Chicago Press, 1959.
Parker, William H. "The Confederate States Navy." In *Confederate Military History*, ed. Charles A. Evans. Atlanta: n.p., 1899.
———. *Recollections of a Naval Officer, 1841–1865*. New York: Scribner's, 1883.
Parkinson, Roger. *The Encyclopedia of Modern War*. New York: Stein & Day, 1977.
Perry, James M. *A Bohemian Brigade: The Civil War Correspondents*. New York: Wiley, 2000.
Peterson, Harold Leslie. *Notes on Ordnance of the Civil War*. Washington: American Ordnance Association, 1959.
*The Photographic History of the Civil War*. Edited by F[rancis] T. Miller et al. 10 vols. New York: Review of Reviews, 1912.
Porter, David D. *The Naval History of the Civil War*. New York: Sherman, 1886.
Ramsdell, Charles W. *Behind the Lines in the Southern Confederacy*. Baton Rouge: Louisiana State University Press, 1944.
Reed, Rowena. *Combined Operations in the Civil War*. Annapolis, Md.: Naval Institute Press, 1978.
Reid, William. *Arms through the Ages*. New York: Harper & Row, 1976.
Robinson, Willard B. *American Forts: Architectural Form and Function*. Urbana: University of Illinois Press, and the Amon Carter Museum of Western Art, Fort Worth, Tx., 1977.
Rosengarten, J. G. *The German Soldier in the Wars of the United States*. Philadelphia: Lippincott, 1886.
Russell, Sir William Howard. *My Diary North and South*. 2 vols. London: Bradbury & Evans, 1863.
Sala, George Augustus. *My Diary in America in the Midst of War*. 2 vols. London: Tinsley, 1865.

Sanborn, Margaret. *Robert E. Lee.* 2 vols. Philadelphia: Lippincott, 1966–1967.

Scharf, J. Thomas. *History of the Confederate States Navy.* New York: Rogers & Sherwood, 1887. Reprint. New York: Fairfax, 1977.

Scott, Henry L. *Military Dictionary.* New York: Van Nostrand, 1861. Reprint. West Point Military Library. New York: Greenwood Press, 1968.

Silverstone, Paul H. *Warships of the Civil War Navies.* Annapolis, Md.: Naval Institute Press, 1989.

Smith, Edward Conrad. *The Borderland in the Civil War.* New York: Macmillan, 1927.

Soley, James Russell. *The Blockade and the Cruisers.* New York: Scribner's, 1883.

Steffen, Randy. *The Horse Soldier, 1777–1978.* 3 vols. Norman: University of Oklahoma Press, 1977.

Still, William N. et al. *Raiders and Blockaders: The American Civil War Afloat.* Washington, D.C.: Brassey's, 1998.

Tarassuk, Leonid, and Claude Blair. *The Complete Encyclopedia of Arms and Weapons.* New York: Simon & Schuster, 1982.

Tousard, Louis de. *American Artillerist's Companion.* 2 vols and 1 vol. of plates. Philadelphia: Conrad, 1809. Reprint. West Point Military Library. New York: Greenwood, 1969.

Upton, Emory. *Infantry Tactics.* Rev. ed. New York: Appleton, 1874. Reprint. West Point Military Library. New York: Greenwood, 1968.

U.S. Congress. *Report of the Joint Committee on the Conduct of the War . . . Second Session Thirty-eighth Congress.* 3 vols. Washington: Government Printing Office, 1865.

U.S. Surgeon General's Office. *The Medical and Surgical History of the War of the Rebellion.* Prepared under the direction of Joseph K. Barnes, Surgeon General. 3 vols in 6. Washington: Government Printing Office, 1870–1888.

U.S. War Department Artillery Officers' Board. *Instruction for Field Artillery.* Philadelphia: Lippincott, 1861. Reprint. West Point Military Library. New York: Greenwood, 1968.

U.S. War Department. *Atlas to Accompany the Official Records of the Union and Confederate Armies.* Washington: Government Printing Office, 1891–1895. Reprint. *The Official Atlas of the Civil War.* New York: Yoseloff, 1958.

———. *Official Records of the Union and Confederate Navies in the*

*War of the Rebellion*. 30 vols. in 2 series. Washington: Government Printing Office, 1894–1927. Vols. 22–27: Western Waters.

———. *The War of the Rebellion: A Compilation of the Official Records of the Union and Confederate Armies.* . . . 70 vols in 129 in 4 series plus atlas. Washington: Government Printing Office, 1880–1901.

Vandiver, Frank E. *Mighty Stonewall*. New York: McGraw-Hill, 1957.

Wagner, Arthur L. *Organization and Tactics*. New York: Westermann, 1895.

Weigley, Russell F. *The American Way of War*. New York: Macmillan, 1973.

———. *History of the United States Army*. Bloomington: Indiana University Press, 1984.

White, William Chapman, and Ruth White. *Tin Can on a Shingle*. New York: Dutton, 1957.

Wilcox, Cadmus Marcellus. *Rifles and Rifle Practice: An Elementary Treatise . . . with Descriptions of the Infantry Rifles.* . . . New York: Van Nostrand, 1859.

Wilson, H. W. *Ironclads in Action: A Sketch of Naval Warfare, 1855–1895*. London: Sampson, Low, Marston, 1896.

**Articles**

"Editorial Paragraphs." *SHSP* 2 (November 1876): 252–53.

"Editorial Paragraphs: Book Notices." *SHSP* 8 (January 1880): 47–48.

"Editorial Paragraphs. *SHSP* 8 (April 1880): 189–90.

Luvaas, Jay. "A Prussian Observer [Scheibert] with Lee." *Military Affairs* 21 (fall 1957): 105–17.

Pierson, William W., Jr. "The Committee on the Conduct of the Civil War." *American Historical Review* 23 (April 1918): 550–76.

Poindexter, Charles. "Major J. Scheibert (of the Prussian Army) on Confederate History: A Review [of a series of articles on the Confederacy]." *SHSP* 18 (December 1890): 422–28.

Russell, Sir William Howard. "Recollections of the Civil War." *North American Review* 166 (February, March, April, May, June 1898): 234–49, 362–73, 491–502, 618–30, 740–50.

"The Siege of Charleston." *Harper's Weekly* 8, 367 (January 9, 1864): 17 (report), 28 (picture).

Venable, C. S. "Major Scheibert's Book [a review of *Bürgerkrieg* in French translation]." *SHSP* 4 (August 1877): 88–91.

Vizetelly, Frank. "Charleston under Fire." *Cornhill Magazine* 10 (July 1864): 90–110.

# INDEX

Abatis: in fortification, 96; and entanglements, 98; in military science, 98
*A. D. Vance* (ship), 121
Adolphus, U.S. Assistant Surgeon: quoted, 108
*Alabama* (ship), 129
*Albatross* (boat), 177
Albermarle Sound, 138
Alcoholic beverages: in Confederate army, 195
Aldie, Va.: skirmish at, 144, 212*n*5
Alexandria, La., 181
Ambulances, 109
Amelia Courthouse: Lee at, 211*n*14
Americans: as military pioneers, 101; attitudes to the war, 118; tendency to exaggerate, 121, 129
Anaconda plan, 20
*Anna* (boat), 202
Antietam (Sharpsburg): campaign and battle, 140–42, 211*n*7
*Arkansas* (boat), 165
Armament: abatis and entanglements, 98; fascines, 99; on Mississippi river, 161–62, 176, 177, 178. *See also* forts and fortresses by name
Arms and weapons: at war's start, 35; Union and Confederate compared, 58; sabers and swords, 61; torpedoes (mines), 87–98, 217*n*3; rams, 123, 124; floating rockets, 217*n*3. *See also* Artillery and ordnance; Firearms
Army, Confederate: mobility, 17; contrasted with Union, 17, 18; after Bull Run, 36; early organization, 36–37; early advantages, 40; Scheibert's estimate of, 40; battle formations, 41; in battle, 41–42; lack of reserves, 41–42; inferiority in numbers, 43, 44, 45, 48, 53, 58; early in war, 43, 44; firearms, 58; health, 100; 108; food and diet, 108; morale, 149–50; attitudes east v. west, 183, 184; leadership, 184; preparation for soldiery, 192, 193, 195, 196; reasons for successes, 192–93, 200; qualities of character, 193–97 *passim;* superior qualities, 195; morality in, 195, 196; character of officers, 196–97; discipline, 196–97
Army, Union: mobility, 17; contrasted with Confederate, 17, 18; early organization, 36; after First Bull Run, 37, 43; short-term-service troops, 38–39; first campaign, 40; battle formations, 41; superiority in numbers, 45, 53, 58; failure in attack, 46; firearms, 58; morale, 185; in West, 185; leadership, 188; advantages, 193
Army of Northern Virginia: morality in, 193, 197–98; importance, 199
Artillery and ordnance: number of Union batteries, 79; smoothbore v. rifled, 79, 80–81; manufacture of Confederate, 81; Lee on, 81; at Chancellorsville, 82; at Gettysburg, 83; American evaluated, 83; Union and Confederate compared, 83; rifled cannon in siege, 83; kinds and makes, 84, 85, 86; manufacture of, 84–86; naval, 86, 92, 125, 130–31, 155, 157–58, 161, 202, 203; American, 88; Americans as artillerists, 88; Union and Confederate compared on service of, 88–89; importance in Civil War arms, 91; at Fort Sumter, 95; at Fort Wagner, 98–99; Union, 140; in war in West, 155–91 *passim;* armament against naval, 161–62, 176, 177, 178; naval in attack on Vicksburg, 164–80 *passim;* in Vicksburg campaign,

239

178, 179, 186; quality of Union naval, 186; naval in combined operations, 190–91; described, 215–16n1, 217n5; measurements used in, xiv, 80, 216n2; carriages in, 216n3; authorities cited, 216–17nn5, 8, 10. *See also* Swamp Angel; kinds by name
Ashby, Turner, 23, 61, 136
Atlanta: battle of, 19, 20
Atlanta, Ga.: factories in, 81
*Atlanta* (ship), 86
Attitudes, values, and character: as martial qualities, 193

Bacon, U.S. Assistant Surgeon: quoted, 108
Bahamas: in blockade running, 120, 121; Nassau, 120; Saint George, 121
Bailey, Theodorus, 128
Banks, Nathaniel P., 22, 23, 129, 181, 182, 183
Banks's Ford, 143
Barges, 161–62
Base of operations: as term and concept, 17–18, 211n2
Baton Rouge, La., 164, 165, 181
Battery Gregg, 95
Battery Wagner, 146
Battle formations, 41–42
Battles, skirmishes, engagements, and raids. *See* each by name
Battles of 1862 and 1863: conduct of, 42
Beauregard, P. G. T., 103, 136, 149, 164, 184
Becker, Colonel Randolph, 202
Belfort, France: fortifications at, 100
Bell, Tyree H., 201
*Benton* (boat), 177, 178, 180; at Grand Gulf, 222–23
Bermuda Hundred, 148
Billinghurst Requa Battery (gun), 217n5
Birds Point, Mo., 156
Bisland, Miss., 181
Black River Bridge, Miss.: skirmish at, 180
Blakeley guns, 85
Blenker, Louis, 24, 198, 224n4

Blockade: running, 120–21, 128–29, 138; importance and effectiveness of, 121–22, 129, 136, 138
Blücher, Gerhard Liebrecht von, 3, 209n5
Boats and ships: Union fleet on Mississippi, 164, 182; construction, 124, 219n6 (chap. 8); barges as gunboats, 161–62; described, 226n3. *See also* each by name
Borcke, Heros von. *See* Von Borcke, Heros
Border states, 137, 155–56, 211n7, 219nn1, 5
Bowen, John S., 178
Bowling Green, Ky., 156
Bragg, Braxton, 30, 167, 186, 212n16
Brandt, Heinrich von: on morality in warfare, 200, 225n8
Brandy Station, Va.: battle of, 65, 66–68, 144, 220n13
Breckinridge, John C., 148, 165–66, 181
Brigade: as tactical unity, 213n7
Bristoe, Va. (campaign), 220n14
Brooke, John M., 122
Brooke gun, 84, 85, 87
Brown, George, 176
Buchanan, Franklin, 123
Buckner, Simon B., 158
Bucktail Rifles, 70
Buell, Don Carlos, 167, 186
Buford, Abraham, 201, 202
Bull Run (battles): First, 19, 22, 35, 36–40, 81, 136; Second, 43, 140–41, 211n5
Burnside, Ambrose E., 20, 25, 142
Butler, Benjamin F., 126, 198, 224–25nn4, 7
Butte à la Rose, 181

Cairo, Ill., 156
Calcium lights, 90
Calhoun (fort) or Fort Wool, 218n2 (chap. 8)
Campion's Hill, Miss.: skirmish at, 180
Canals: in Vicksburg campaign, 171–72

Cannon. *See* Artillery and ordnance; each gun and cannon by kind
Cape Girardeau, Mo., 156
*Carondelet* (boat): against Fort Donelson, 157, 158; against New Madrid, 162, 163; against Vicksburg, 176, 177; in attack, 185 (Island Number 10), 222 (Grand Gulf)
Carriages: artillery, 216$n$3
Casey, Silas: *Infantry Tactics* cited, 56, 214$n$3
Casey's tactics, 56–57
Caswell (fort), 94, 96
Catlett's Station, Va.: Stuart's raid on (Peninsular campaign), 68–71, 140
Cause, the Confederate, 196
Cavalry: foot, 23; Confederate, 30, 136, 190; European, 59; American lessons for European, 59–60, 74; American influence on European, 60; American evaluated, 60; American compared to Prussian and German, 60, 74–77; Union and Confederate compared, 60–61, 74; Union evaluated, 61; arms and weapons, 62; nature and purpose, 63, 65; value to Lee, 63–64; Confederate described and evaluated, 63–64, 65, 74–75, 77; in battle, 64–65; against artillery and infantry, 64–65, 214$n$1; qualities needed in officers and men, 76; Prussian and German manual of procedures, 76–77; training and drill, 77; need for engineers, 78, 215$n$14; Confederate need for, 179; dismounted, 214$n$6; deployment of, 215$n$9; size of Stuart's, 215$n$10
*Cayuga* (ship), 128
Cedar Run, Va. (creek), 140
Cemetery Hill: in Battle of Gettysburg, 144
Chalmers, James R., 202
Chancellorsville (battle), 26–28, 42, 43–44, 49, 142–43; Lee at, 49; artillery at, 82; casualties, 106–7, 108; care of wounded at, 106–8; importance, 192; Pleasonton's testimony on, 211$n$9
Chanzy, Antoine (French general), 78

Charleston, S.C.: siege and battle of, 84, 135, 145–46; artillery at, 86, 91; in blockade, 120; bombardment of, 216–17$n$8
Chattanooga, Tenn.: campaign and battle, 149, 167, 184, 212$n$16
Chesney, Charles Cornwallis: book on U.S. Civil War cited, 60, 215$n$3
Chickahominy (Gaines's Mill), Va.: Peninsular campaign, 139, 140
Chickamauga, Tenn.: campaign and battle, 149, 212$n$16
Christianity: as source of morality, 195, 200
*Cincinnati* (boat), 157
Civil War (U.S.): lessons for Europe, xi, 34, 38, 86, 93, 100, 102 (pioneering), 105 (medical), 114–16, 119, 122, 124 (naval), 130, 136, 212–13$n$1, 214$n$2; as textbook and lesson for warfare, 13, 92–93; how fought, 34; beginning of, 35; early conduct of, 35–37; artillery developed in, 91; North, South, and West compared, 92; engineering developed in, 93; consequences in West, 117, 189; attitudes to, 135, 212–13$n$1; causes of, 135–36; historical importance, 189, 192
Coehorn mortars, 99
Cold Harbor: First (Gaines's Mill), 32, 43, 139, 140; Second, 47, 148; seriousness of Second, 220$n$17
Cold Spring foundry (or West Point foundry), 79, 88
Columbia, S.C.: factories in, 81
Columbiad (cannon), 84–87 *passim*, 95, 125, 179
Columbus, Ky., 156, 159, 184
Combined operations: defined and evaluated, 92, 126, 151–53, 185, 189–91, 220 (headnote, chap. 10); importance of, 126, 186, 188, 189; at Island Number 10, 185; at Vicksburg, 187–88; navy in, 187–88; attitudes to, 189–91; summary of chapter on, 189–91; in Mississippi theater, 191; need for, 191
Committee on the Conduct of the War: report quoted, 54–56, 199

Communications: in infantry attacks, 41; scouts and couriers for, 62, 63, 67, 72, 75; newspapers for information, 63; cavalry useful in, 63–64; Stuart's sources of information, 64; importance of, 75; along Mississippi, 183; communiques and letters, 197; in Mississippi valley campaign, 221n6. See also Telegraph
*Conestoga* (boat), 157, 226n4
Confederacy, the: attitudes in, 141, 146–47; situation in 1863, 143; advantages and strengths, 147; Scheibert's favor of, 192, 219n8; belief in the Cause, 196; Prussian attitude to, 209n1. See also Army, Confederate; South, the; Union v. Confederacy
*Congress* (ship), 123
Conyingham, David P.: quoted on Sherman's Georgia march, 198–99
Corinth, Miss., 160, 186; battle, 137, 142, 167
Couriers: cavalry, 62, 72, 75; as messengers at Brandy Station, 67; outfitting of, 215n5
Crimean War: lessons of, 214n1, 217n2
Cromwell, Oliver, 196
*Cumberland* (ship), 123

Dahlgren, John A., 145–46
Dahlgren (gun), 87, 125, 179
*David* (boat), 217n4
Davis, Jefferson, 136, 144, 187
Dawson, Samuel K., 203
De Russey (fort), 181
DeShroons, Miss., 180
*DeSoto* (boat), 176, 182
Diet: of the armies, 17; Confederate, 108
Dismounted cavalry, 214n6
Dix, John, 25, 30
Dodge, Grenville M., 168
Donelson (fort), 93, 137, 156, 159, 184
Dover, Tenn., 158
DuPont, Samuel F., 145

Early, Jubal A., 198
East, the: as theater of war, 138–40

Echelons: in cavalry formations, 65
Eighth New York Militia: volunteer artillery, 39
Ellet, Charles R., 176
Elliot, Stephen, 103
Emancipation Proclamation, 142
Engineers and engineering, 140, 185; in cavalry, 78, 215n14; Union v. Confederate, 93; pioneers in, 101; canal digging, 171–72
Entanglements (fortification), 96, 99
*Essex* (boat), 182
Ethics. See Morality
Ewell, Richard S., 28, 29–30, 212n14

Faidherbe, Louis (French general), 78
Fair Oaks (Seven Pines): battle of, 139, 211n3
Fairview Hill, 106
Farragut, David G., 124, 177; at New Orleans, 126, 127, 138; in West, 140; evaluated, 140, 185, 186; reconnaissance of Mississippi River, 163–64, 176; in attack on Vicksburg, 164–80 *passim*, 185; and canals, 171; at Fort Hudson, 182; defies torpedoes, 219n4 (chap. 8); on Vicksburg defenses, 221n9
Fascines (fortification), 99
Ferraro, Edward, 103
Firearms: Union and Confederate compared, 35. See also Arms and weapons
Fisher (fort), 87
Flags, banners, and guidons: hospital, 81, 109; battle flag defined, 215n11
Fletcher, Henry Charles, 200
*Flora* (ship), 126
*Florida* (ship), 128
Floyd, John B., 158, 221n7
Folly Island, S.C., 98, 146
Food and provisions, 17
Foot cavalry, 23
Foote, Andrew H., 157, 160, 161, 220n1
*Forest Queen* (boat), 177
Forrest, Nathan Bedford, 168, 201–4 *passim*
Forts and fortifications: masonry v. earth, 90, 95–96, 100–101; theory of,

93, 94, 217n1 (chap. 6); in South, 93–94; described, 94; abatis in, 96; entanglements in, 96, 99; field, 96–97; underwater nets, 97; at Fort Wagner, 98; parallels in, 98, 99; in military science, 98, 189; fascines in, 99. *See also* forts by name
Fourth Pennsylvania Volunteers, 39
Franco-Prussian War, 48
Frayser's Farm (White Oak Swamp) skirmish at, 139, 211n6
Frederick the Great: as model soldier and commander, 21, 33, 41, 51–52
Fredericksburg (campaign and battle), 44; Lee in, 172; Burnside in, 220n12; Hooker in, 220n12
Frémont, John C., 22, 139, 210n9

Gaines's Mill (Chickahominy or First Cold Harbor), 139
Gardner, Franklin, 178, 183
Generalship: early in war, 18; vagabond style of, 18; nature in this war, 18–19; early failure of Union, 20; necessities for good, 26–27
*Genessee* (boat), 182
Geography and terrain: as military factors, 13–15, 19, 20, 46, 78, 92, 96, 100, 101, 107; influence on tactics, 13–15, 20, 46, 48, 59–65 *passim*, 74, 78, 80, 82, 92, 138, 139, 145, 153–55, 181, 187; influence on cavalry, 59–60, 63, 65, 74, 82; influence on artillery, 80; of North, South, and West compared, 92–93; influence on strategy, 153–55, in war in West, 159; influences on Mississippi defenses, 181, 187
Gettysburg (campaign and battle), 28–31; Confederate objectives, 28, 29; consequences, 30–31, 44–45, 144–45; artillery at, 82–83
Gibbon, John: quoted, 54–56, 213–14n2
Gillmore, Quincy A., 90, 98, 145–46
Gneisenau, August, Graf Neithardt von: Prussian strategist, 209n5
Gracy, Captain, 203
Grand Gulf, Miss.: as fortified village, 179

Grand Gulf, Miss. (fort), 166, 180; importance of, 191; plan for attack on, 222–23; Union navy order of battle at, 222–23
Grand Junction, Miss., 167
Grand Lake, Miss., 181
Grant, Ulysses S.: character of, 5–6, 188; as commander, 5–6, 19, 31–32, 45, 47, 51, 188–89, 204; evaluated, 5–6, 140, 145, 188–89; campaigns, 6; life and career, 6, 188; as strategist, 20, 32; and Lincoln adminstration, 29; v. Lee, 31–32, 47–48, 51, 147–50 *passim*; at Battle of Corinth, 137; in West, 140; takes Vicksburg, 145, 179, 188, 189; as general-in-chief, 145–50 *passim*; in command of West, 167–80 *passim*; in personal command of Mississippi theater, 170–71; orders Ross to Helena, 172; recalls Quimby, 174; at New Carthage, 176; and Port Hudson, 183; public opinion on, 188; in attack on Vicksburg, 188; use of combined operations, 188, 189, 190–91; and naval support, 190–91; relations with Sherman, 197; in orders of battle, 204; and Frémont, 209n9; and liquor, 210n8; at war's start, 210n9; reappointed to Union general, 210n9; and Elihu B. Washburn, 210n9; as president, 6, 210n10; and Petersburg, 212n17; and unconditional surrender, 221n7
Gray, A. B., 161
Gierson, Benjamin H., 179
Grover, Cuvier, 181
Gregg (battery), 95
Grenada, La., 178
Guns. *See* Arms and weapons; Artillery and ordnance; Firearms; kinds of artillery by name
Gustavus, Adolphus, 196

Haines Bluff, Miss., 180
Halleck, Henry, 161, 184, 186; evaluated, 166–67, 186
Hamilton, Schuyler, 160
Hampton, Wade, 71, 74
Hardee, William Joseph: his *Rifle*

*and Light Infantry Tactics* cited, 57, 214n5
Hard Times, La., 176, 178, 179, 180
Harpers Ferry, Va.: arsenal and armory at, 136, 141, 219n2; Jackson at, 141, 209n2; described, 209n2; mission of Confederates at, 209n2
Harrison's Landing, Va., 139
*Hartford* (ship), 125, 128, 177, 182
Hatteras (fort), 137–38
Health and fitness: Confederate army, 108
Heiman, Adolphus, 157
Heiman (fort), 201, 202, 213
Helena, Ark., 183
Henry (fort): importance, 93, 137, 156, 183, 191
*Henry Clay* (boat), 177, 178
Hill, Ambrose P., 3, 27
Hill, Daniel H., 25, 143
Hoel, William R., 177
Hoke, Robert F., 146
Holly Springs, Miss., 169, 187
Hood, John Bell, 149
Hooker, Joseph, 20; as commander, 26–28; at Chancellorsville, 26–28, 43–44, 142–43, 211n9, 212n11; as commander of Army of the Potomac, 142–43; nickname, 142, 219–20n11; size of forces, 192
Hospitals. *See* Medical service
Hunter, David: behavior of troops under, 198; destruction by, 224n4, 225n7; burns Virginia Military Institute, 224n4
Hurlbut, Stephen A., 170, 171, 204

*Indianola* (boat), 176
Industry: manufacture of artillery, 79, 80, 84–86; North v. South in, 92
Interior lines, 21–25 *passim*, 28, 31, 211n3
Ironclads, 124
*Ironsides* (ship), 97
Island Number 10: importance, 137, 156, 159, 184, 185; attack on, 160, 161; in attack on New Madrid, 162; Confederate loss of, 163, 190; combined operations at, 185
Iuka, Miss.: battle of, 167

*J. W. Chaseman* (boat), 202, 203
Jackson (fort), 93, 127
Jackson, Miss.: skirmish at, 179, 180
Jackson, Stonewall: character, career, and appearance, 2–3, 196; attitude of men toward, 2, 3, 196; evaluated, 2–3, 24, 42, 44; as commander in early action, 2–3, 20, 22–25 *passim*, 68, 79, 139, 140; death of, 3, 143, 209n4; last words of, 3, 209n4; use of interior lines, 22–24; in the Shenandoah, 42, 139, 192; at Chancellorsville, 49, 82, 106, 143; at First Bull Run, 136; in Second Bull Run campaign, 140; at First Cold Harbor, 140, 148; at Harpers Ferry, 141, 209n2
Jacobi, S.: cited on artillery, 86, 216n8
James Island, 97
James-York peninsula, 138
Johnson (fort), 97
Johnston, Albert Sidney, 137
Johnston, Joseph E.: in Peninsular campaign, 22; and interior lines, 22, 211n3; against Sherman, 149; against Grant, 168, 171; and Vicksburg, 178; evaluated, 184; as commandant of West, 187; at Fair Oaks, 211n3

Kanawha (river), 137
*Kearsarge* (ship), 129
Kelly's Ford, Va., 26
Kenly, John R., 23
Kernstown, Va.: First Battle of, 23
*Kineo* (boat), 182

*Lafayette* (boat), 177, 222
Lake Providence canal project, 171, 172, 188
*Lancaster* (boat), 177
Ledlie, James H., 103
Lee, Fitzhugh, 26, 71, 74
Lee, Robert E.: as commander, xii, xvi, 8–9, 17, 21, 25–32 *passim*, 44, 46, 47, 51, 52, 146; character of, 6–9 *passim*, 42, 52, 150, 194–96 *passim*; evaluated, 6–9 *passim*, 32, 44, 47, 150, 192, 194; as strategist, 7, 21, 32; appearance, 7, 8; life and career, 7,

Index    **245**

9; lifestyle, 8–9; headquarters, 9; death, 9; Scheibert's relationship with, xi, 9, 209n6; and Army of Northern Virginia, 25; at Chancellorsville, 26–28, 42, 43–44, 143; strategy before Gettysburg, 29; against Grant, 31–32, 47–48, 51, 147–50; success as campaigner, 32; on command in battle, 42; at Cold Harbor, 43; at Antietam, 44, 141; at Fredericksburg, 44, 142; at Gettysburg, 45, 82; on Prussian army, 49; at war's end, 52; and Brandy Station, 66; on artillery, 81, 82; and Petersburg mine assault, 103; in western Virginia, 137; and battles of 1862, 139; in Maryland, 141; against Hooker, 143; last campaigns and battles, 147–50 *passim*; at Amelia Courthouse, 150; surrender, 150; threatens Washington, 167; size of forces, 192; march to Gettysburg, 194–95; desire for morality in forces, 194–95; General Order 73 quoted, 194–95; attitude of men toward, 196; letter to Jackson quoted, 209n6; later life, 210n11; marriage and personal life, 210n11; and Confederacy, 210n12; as educator, 211n15; and Petersburg, 212n17; and cavalry at Gettysburg, 215n8

Lee, Rooney, 71
Lee, Samuel P., 164
Lee family, 7, 210n11
*Lexington* (boat), 167, 226n4
Lexington, Miss.: engagement at, 168
Lighthouse Creek, 99
Lincoln, Abraham: as commander in chief, 29; opposes Hooker about Gettysburg, 29; government of, 29, 188, 212n12, 219n8; election of, 135; calls for volunteers, 136; and Emancipation Proclamation, 142
Liquor: in Confederate army, 195
Logistics, 17–18, 19
Longstreet, James, 25, 31, 197, 220n15
Looting, 197–200, 213, 224n4
Loring, William W., 171, 173, 178
*Louisiana* (boat), 127

*Louisville* (boat), 158, 177, 222
Louisville, Ky., 167
Lowe, William W., 168
Lyon, Hylan B., 201

McCausland, John, 224n3
McClellan, George B.: use of navy, 19–20; strategy of, 19–20; as commander, 20, 43, 141, 142; mistakes, 20; in Peninsular campaign, 22; evaluated, 39–40; organizes Union army, 39–40, 219n4; and artillery, 79; and blockade, 122; in western Virginia, 137; early organization, strategy, and leadership in battle, 138–39; against Jackson, 140; at Antietam, 141; at Cold Harbor, 148; morality of forces under, 200
McClellan saddle, 62
McClernand, John A., 167, 179, 180, 186; as co-commandant on Mississippi, 170, 188; in orders of battle, 204
McCown, John P., 160, 161
McDowell, Irvin: official report quoted, 38–39, 136, 139–40
MacKall, William W., 161, 163
Macon (fort), 138
McParlin, Thomas A.: quoted on medical care, 109–13
McPherson, James B., 170, 171, 204
Magruder, John B., 138
Mahone, William, 103
Malvern Hill, Va.: engagement at, 139, 211n6
*Manassas* (ship), 127, 128
Manassas. *See* Bull Run, First
March discipline: Confederate, 57–58
Maryland: attitudes as Border State, 141
Massaponax Church (Thornburg): skirmish at, 211n6
*Mazeppa* (boat), 201, 202
Meade, George G.: as McClellan's successor, 20, 146; in Gettysburg campaign, 144; after Gettysburg, 45, 145; morality of forces under, 200
Mechanicsville: battle of, 139

Medical service: ambulance, 106, 109; equipment and supplies, 106; care of wounded, 106, 108–11; surgeons, 106, 110–11, 115; dressing stations and field hospitals, 106, 107, 109–11; transporting sick and wounded, 111–13; in Wilderness, 111–13; Union and Confederate compared, 113–14; hospitals listed, 114; American described, 114–16; attitudes toward wounded over history, 115–16

Meerheimb, F. von: quoted on Sherman, 3–5, 210$n$7

Memphis, Tenn., 137, 140, 184

*Merrimac* (ship), 122–23, 124, 130, 226$n$4. See also *Monitor* v. *Merrimac*

Middleburg: engagement at, 144, 212$n$15

Military literature: phrasemongery in, 193–94

Military science: American and German compared, 48; history v. theory in, 59; artillery in, 80–84 *passim*, 91; theory v. practice in artillery, 83–84; sieges, 98; engineering, 100–101; ironclad ships, 124, 130–31; combined operations, 126, 151–53, 185; fortifications and fortresses, 189; lessons for, 189

Milliken's Bend canal project, 175–76, 178, 188

Milroy, Robert H., 23, 28, 224$n$4

Mine Run: campaign, 220$n$14

*Minnesota* (ship), 123

*Mississippi* (ship), 128, 182

Mississippi River: Union navy on, 137, 164; fighting along, 140, 142; combined operations on, 153–55; importance of, 163, 181, 183, 185; Union takes control of, 183

Mississippi River system: war in West on, 154–91 *passim*

Mississippi theater: hardships in, 191

Mississippi valley: importance of, 153–55, 186, 188, 189; Confederate attitute toward, 184; lessons for combined operations, 189. See also West, the; Vicksburg campaign

Mobile, Ala., 164

Mobile Bay: battle of, 86, 97, 124–26

Mobility in warfare, 211$n$2

Moltke, Helmut Karl Bernhard, Graf von, 42, 213$n$4

*Monitor* (ship), 89, 226$n$4; v. *Merrimac*, 89, 123–24, 126, 138

*Monogahela* (boat), 148

Monroe (fort), 138

Morality: and attitude as martial qualities, 193, 194; importance, 193–94; in military science and warfare, 193–200 *passim;* in Southern forces, 194–95, 196

Morgan (fort), 124, 125

Morgan, George W., 169

Morgan, John H., 197

*Morgan* (ship), 127, 128

Morris Island, S.C., 146

Mosby, John S., 62, 63, 215$nn$7, 9

*Mound City* (boat), 177, 222

Mud march: as term, 211$n$8

Murphy, J. McCleod, 177

Murphy, R. C., 169

Napoleon cannon, 80

Nassau, Bahamas: in blockade running, 120

Natchez, Miss., 164, 166

Naval development: New England and, 117–18

Naval issues: ironclad v. wood, 124

Naval warfare, 117, 130–31

Navies: importance in warfare, 117

Navy, Confederate: in West, 207

Navy, Union: McClellan's use of, 19–20; superiority of, 92, 191; statistics on ships and sailors, 118; strength, 129–30; at Malvern Hill, 139; in West, 140, 185, 186; importance, 153–54; western fleet, 155–91 *passim*, 169, 177–78, 205–6, 207; Confederate attitude to, 178; in attack on Vicksburg, 179; damage to, 182; at Fort Donelson, 184; in combined operations, 187–88; gunboats in, 190

Negroes and Negro troops, 93, 98, 103, 104, 165

Nets: as harbor defenses, 97

Index  247

New Carthage, La., 176
*New Ironsides* (boat), 217n4
New Madrid, Mo., 137, 156, 159–60, 161–63, 184, 190
New Orleans, La., 126–28, 164, 165
Newspapers, 63
Nichols, George Ward: cited on Sherman's march, 199, 224n5
Norfolk, Va., 136, 138–39; as naval base, 219n2
North, the: industries, 79, 84; manpower and resources, 147, 149–50; advantages, 192, 193; superior qualities of, 193; care for sick and wounded in, 200; morality in, 200. *See also* North v. South; Union
North Anna: battle of, 147
North v. South: compared on major factors relative to war, 84–85, 92, 192–94; attitudes, talents, and behaviors, 100; as engineers and technicians, 100; aptitude for pioneering, 101; care of wounded, 105–6, 107–8; Union treatment of Confederate wounded, 108; medical services, 113–14; hatred on both sides, 135; relative strengths, 136, 137, 140, 142–43; as sources of soldiers, 193. *See also* Confederacy; Union
North v. South and West: character, attitudes, and abilities in, 92

Officers: necessary characteristics, 35–36, 37, 38, 51, 200; attitudes of Confederate, 37; Americans as commanders, 48–49; American and Prussian compared, 49–50; qualifications of American, 50; how chosen, 53, 54–56; quality of Confederate, 53, 54; quality of Union, 54–56; training of, 54–56; qualities needed for cavalry, 75; branches chosen by prominent, 79; behavior of, 200
*Oneida* (boat), 128, 164
Orders of battle: Union and Confederacy, 204–5

Organization of army: French example, 56–57
Owen, Elias K., 177

Pace: as unit of measurement, xiv, 216n2
Paine, Eleazer A., 163
Pamlico Sound, 138
Parallels: in fortifications, 98, 99
Paris Landing, Tenn., 201, 202
Parkman, Francis: quoted on officers, 213n1
Parrott, Robert C.: foundry of, 79
Parrott gun, 80, 87–88, 201; Swamp Angel, 99
Partisans: cavalry, 62
Patterson, Robert, 22, 39
Pemberton, John C., 167, 168, 171, 173, 186; surrenders Vicksburg, 180–81, 189; in orders of battle, 204–5
Pemberton (fort), 178
Peninsular campaign, 22; plunder by Union forces in, 200
Petersburg: campaign, assault, and siege, 102–4, 148–49, 212nn17, 18
Pickett, George E., 146
Pillow (fort), 156
*Pittsburg* (boat), 160, 177, 222
Pittsburg Landing (Shiloh): battle of, 137
Plank roads: as term for corduroy roads, 14, 211n1
Pleasonton, Alfred, 26, 29, 30, 212n14
Point Pleasant, La., 160
Polk, Leonidas, 156
Pope, John, 20, 24–25, 186; at Battle of Second Bull Run, 43; headquarters looted, 69–70, 71; as commander in Virginia, 140; evaluated, 145; in West, 159; looting by, 224n7
Porter, David: evaluated, 140; and Vicksburg, 16, 169, 170, 172, 175, 177; letter by, quoted, 179; at Arkansas Post, 220n1; General Order of, quoted, 221–23n18; at Grand Gulf, 222–23
Porter, William D., 157
Port Gibson, Miss., 180

Port Hudson, La., 166, 178, 181, 182, 183
Potter, Joseph, 103
Price, Sterling, 167, 186
Pulanski (fort), 95, 138

*Queen of the West* (boat), 176
Quimby, Isaac F., 173–74
Quinine, 100

Railroads: importance of, 15–16, 149, 169, 186; used by Lee, 25; and Grant before Richmond, 148; troops moved by, 168; compared to boats, 190
Rams, 123, 124
Randolph (fort), 156
Ransom, Robert, 146
Raymond, Miss.: skirmish at, 180
Red River, 181, 182
Regulars: as cavalry, 62
Religion: importance to soldiery of, 195–96
*Report of the Committee on the Conduct of the War*: quoted, 54–56, 199
Requa (gun), 98; Billinghurst Requa Battery, 217$n$5
Revolutionary War (U.S.), 33
Rhett, Alfred A., 90, 217$n$10
Richmond, Va.: importance of, 19, 24, 25, 29, 31, 32, 135, 136, 138, 139, 150, 211$nn$4, 5; as Union objective, 19, 25, 31, 32, 40, 43, 139–40, 148–49; during Gettysburg campaign, 29; as industrial center, 81, 84
*Richmond* (boat), 182
Roanoke Island, N.C., 137–38
Robertson, Beverly H., 66
Rockets, floating, 97, 217$n$3
Rodes, Robert Emmet, 212$n$14
Rodman method of casting artillery, 86, 87
Roerdansz, Prince Hohenlohe: on artillery, 83
Rose, Leonard F., 172, 173
Rosecrans, William S., 167
Rosser, Thomas L., 70

Sable Island, La., 127
*Sachem* (boat), 182

*St. Louis* (boat), 157, 158
St. Philip (fort), 93, 127
Scheibert, Justus: and Bismarck, xi; in Prussian army, xi; and Robert E. Lee, xi, 209$n$6; in Confederacy, xi, xii; as military observer and authority, xi–xii; as authority on U.S. Civil War, xii; purpose, method, and point of view in *A Prussian Observes the American Civil War*, xi–xii, xv, 1, 6, 13, 21, 28, 60, 61, 219$n$6; as eyewitness, 40, 42, 49, 57, 58, 61, 85, 89, 93, 192–97 *passim*, 209$n$6; pamphlet on artillery by, 83; at James Island, 89; and Confederate engineering, 93; as judge of U.S. Civil War matters, 193; use of statistics, 211$n$3; standards of judging greatness, 211$n$16; preference for Confederacy, 218$n$1, 219$n$8, 223$n$1; critique of U.S. Civil War books, 220–21$n$5
Schenck, Robert C., 23
Scott, Winfield, 37
Scouts, 62, 75
Sedgwick, John, 26, 143
Sevastopol: siege of, 95, 217$n$2
Seven Pines (Fair Oaks): battle of, 139, 211$n$3
Seven Years' War, 51
Sharpsburg (Antietam): campaign and battle, 140–42, 211$n$7
Sharpshooters and snipers, 65, 67, 99, 160, 162, 202, 215$n$9
Shenandoah Valley, 15, 22, 194, 198
Shepherdstown, W. Va.: skirmish at, 141
Sheridan, Philip H.: behavior of troops under, 198; looting and destruction by, 224$n$4, 225$n$7
Sherman, William T.: life, character, and career of, 4–5; in Atlanta campaign, 19, 20, 32, 149, 150, 192; as artillery officer, 79; in command in West, 149, 204; in Vicksburg campaign and battle, 168, 169, 170, 174, 175, 180, 186, 188; at Walnut Hills, 187; as co-commandant on Mississippi, 188; relations with

Grant, 197; against Early, 198; in orders of battle, 204
Shields, James, 23, 24
Shiloh: battle of, 137, 184
Sickness and medicine, 100. *See also* Medical service
Sigel, Franz, 22-23, 139, 146, 148
*Silver Wave* (boat), 177
Smith, Andrew Jackson, 169
Smith, Martin L., 164, 169
Smith, Morgan L., 169
Snipers. *See* Sharpshooters and snipers
Soldiers: Union v. Confederate, 35-38 *passim*, 42, 53, 57, 58; attitudes of Confederate, 53; nature and quality of Union, 53, 58, 61; Confederate march discipline, 57-58; Confederate cavalrists, 61, 63, 64, 65; need for morality, 194-200 *passim*
South, the: hatred for Yankees, 66; enthusiasm for the Cause, 66; industries, 81, 84; as theater of war, 137-38; early strategy of, 138; attitudes of, 141; manpower and resources of, 147; as source of soldiers, 193. *See also* Confederacy; North v. South
*Soutiens*: French tactical maneuver, 48
Spies, 64, 73
Spotsylvania, Va.: campaign and battle, 47, 147-48
*Star of the West* (ship), 118
Steele, Frederic, 169
Steele's Bayou project, 174-75
Stevenson, Carter L., 171, 178, 204
Strategy: Union and Confederate, 18-26 *passim*; of Confederates, 138-39; mobility in, 211n2. *See also* Tactics
Stuart, J. E. B.: character, career, and appearance, 1-2, 71, 73-74; as raider and cavalry commander, 26, 43, 61-66 *passim*, 71-72, 76; at Brandy Station, 67-68; fame of the "ride around McClellan," 72, 74; and Chancellorsville, 106, 211n9; White House raid, 139; at Second Bull Run, 140; and Gettysburg, 144, 213n5;

cavalry of, 215n10; in Maryland, 219n10
Sullivan, Jeremiah C., 168
Sumter (fort): bombardment of, 85, 89-90, 135, 217n7; artillery at, 89; described, 89-90; as example of masonry fortifications, 95; smashed by Gillmore, 146
*Sumter* (ship), 129
Surgeons and surgery. *See* Medical service
Swamp Angel (gun), 99, 217n6
*Switzerland* (boat), 177

Tactics: in American skirmisher warfare, 33-34; American lessons for Europeans, 34, 48, 50-51; formations in attack, 41; in offense and defense, 43-48, 51; shaped by geography and terrain, 45-46; American, Prussian, and German compared, 48-51; main principles of American, 50-51; artillery, 82-83; mobility in, 211n2; brigade in, 213n7. *See also* Geography and terrain; Strategy
Taylor, Richard, 96, 181
*Tecumseh* (boat), 218-19n4 (chap. 8)
*Tennessee* (ship), 86, 124, 125
Tennessee River, 137
Telegraph, 16-17, 70
Terrain. *See* Geography and terrain
Thornburg (Massaponax Church): skirmish at, 211n6
Tilghman, Lloyd, 156
Tiptonville, La., 160, 161, 163
Torpedoes, 97, 98, 217n3
Tredegar Works: cannon foundry, 84
*Tuscumbia* (boat), 177, 180, 222
*Tyler* (boat), 157, 226n4

*Undine* (boat), 202, 203, 204, 212, 225n1
Union, the: early mismanagement of army and war, 29; boats on Mississippi, 172; attitude to Mississippi valley, 188; manpower, 191, 192, 213n6; citizen fitness for army, 195. *See also* Army, Union; Army, Confederate; North;

South; North v. South; Union v. Confederacy
Union v. Confederacy: attitudes, character, and capabilities, 92–93; size of forces, 146, 149, 192, 213*n*7, 221*n*7; military successes, 191–92; quality of soldiers, 191–98; morale, 193; cavalry, 60, 61, 215*n*4; hospitals, 105, 108–10, 113, 217*n*1 (chap. 7), 218*n*2
Upperville, Va.: engagement at, 144, 212*n*5

Vagabond style of generalship, 18
Van Dorn, Earl, 62, 164, 165–66, 168, 171, 186
*Varuna* (boat), 128
*Venus* (boat), 202, 203, 225*n*1
Vicksburg: campaign and battle, 164–80 *passim*; siege of, 140, 149; importance of, 142, 163, 181, 191; surrender, 145, 183, 191; canal digging, 171–72; navy in, 172, 177–78, 179, 180; siege and attack projects summarized, 175; Milliken's Bend canal, 175–76; Union fleet in attack, 177–78; defenses and fortifications, 178, 179, 185–86, 188; Confederate attitude to Union about, 179; called Fortress of the West, 179; bombardment in attack, 179–80, 187–88; combined operations, 185, 187–88, 191. *See also* Mississippi River; Mississippi theater; Mississippi valley; West, the
Virginia: principal seat of war, 13, 153; geography of, 13–15; roads, 14; as theater of war, 37, 137, 153
*Virginia* (boat), 226*n*4
Virginia Military Institute, 224*n*4
Von Borcke, Heros: quoted, 1; wounded, 30, 212*n*15; at Brandy Station, 65, 66–68; memoirs quoted and cited, 66–71, 72, 215*nn*11, 13; at Catlett's Station, 68–71; as cavalry leader, 71; evaluated, 72; on Stuart as raider, 72–74; attitude to Confederacy, 209*n*1; Southerners' attitude toward, 209*n*1

Wade, James F., 180
Wagner (fort): artillery at, 89, 90–91; attack on, 98; defenses, 98–100; bombardment of, 94–95; described, 94–95, 97, 146; sickness in, 95
Walke, Henry, 162, 177, 185, 222–23
Walnut Hills, Miss., 169, 188
Warfare: art v. science in, 193–94; requirements for success in, 211*n*2. *See also* Military science
Washburn, Elihu B.: and Grant, 210*n*9
Weapons. *See* Arms and weapons
*Webb* (boat), 177
West, the: as theater of war, 92, 136–37, 140, 142, 145, 155, 189–91; importance of, 153–55, 189; strategy in, 156; Union control of Mississippi River, 183; Confederate attitude toward, 183–84; Confederate leadership in, 186–87; behavior of armies in, 197. *See also* Mississippi River; Mississippi theater; Mississippi valley; Vicksburg
West Point, 54
West Point foundry (Cold Spring foundry), 79, 88
White House, Va.: raid on, 101–2; as supply base, 139; on Stuart's ride around McClellan, 217*n*10 (chap. 6)
White Oak Swamp (Frayser's Farm): skirmish at, 139, 211*n*6
Whiting, William H., 146
Whitworth cannon, 80
Wilderness (campaign and battles): Lee v. Grant, 47, 146–48; care of wounded in, 109–13; Chancellorsville in, 143
Willcox, Orlando B., 103
Williams, Thomas, 164, 166
Williamsport, Md., 144
Wilmington, N.C.: in blockade running, 119, 120, 121
Wilson, Byron, 177
Wilson, William, 212*n*1
Wilson's Bayou, Miss.: canal, 161
Wool (fort) (Fort Calhoun), 122, 218*n*2

Yazoo Pass and River: in Vicksburg campaign, 171, 172, 174, 188